Hydrology and the
River Environment

Hydrology and the River Environment

Malcolm Newson

CLARENDON PRESS · OXFORD

Oxford University Press, Walton Street, Oxford OX2 6DP

Oxford New York
Athens Auckland Bangkok Bombay
Calcutta Cape Town Dar es Salaam Delhi
Florence Hong Kong Istanbul Karachi
Kuala Lumpur Madras Madrid Melbourne
Mexico City Nairobi Paris Singapore
Taipei Tokyo Toronto
and associated companies in
Berlin Ibadan

Oxford is a trade mark of Oxford University Press

Published in the United States by
Oxford University Press Inc., New York

First published 1994
Reprinted 1995

British Library Cataloguing in Publication Data
Data available

Library of Congress Cataloging in Publication Data
Newson, Malcolm David.
Hydrology and the river environment/Malcolm Newson.
Includes bibliographical references.
1. Hydrology. 2. Rivers. I. Title.
GB661.2.N49 1994 551.48—dc20 93–12839
ISBN 0-19-874156-1
ISBN 0-19-874157-X pbk

Printed in Great Britain
on acid-free paper by
Bookcraft Ltd
Midsomer Norton, Avon

Acknowledgements

The author acknowledges the key importance of his 'apprenticeship' in hydrology at the UK Natural Environment Research Council's Institute of Hydrology. Under people such as Jim McCulloch, John Sutcliffe, and Mike Lowing I learned very quickly the multi-faceted joys of the subject and deviously planned to add even more facets of my own! Bedding hydrology down nearer the social sciences occurred, however, as the result of thirteen years' physical science on Plynlimon as a field-worker at the Institute's prestigious catchment experiments. Since that study focuses upon the influence of land use on runoff and water quality in the headwaters of two of Britain's most important rivers, it was inevitably socio-political and economic forces which took our results forward, and it was on the back of changes in public attitudes to the environment as a whole that something 'was done' about those results. So, heartfelt thanks to the 'Hydro Boys' up there on the mountain, more especially to the one in Plate 2.3(b) and her daughters.

Finally, working within a Geography department once again has given me all the excuses I need for breadth; careful warnings from both colleagues and students alike remind me that depth need not be sacrificed! In the hope that this book brings hydrology to a wider audience, may I also acknowledge those in planning, development, conservation, farming, and forestry who have given me the chance to demystify the subject for them, often in the pouring rain!

In the production of the manuscript, may I acknowledge the enormous help provided by Sheila Spence on the typescript, by Nick Staley (formerly a 'Hydro Boy' himself) who worked so skilfully and rapidly on the figures, and Ann Rooke. Clare Wake, a hydrology student, completed the index. At Oxford University Press Andrew Schuller kept reminding me of my promise to write this, Enid Barker was very tolerant of my aberrant style, and John Callow worked wonderfully to bring the chaos of the first typescript 'under starter's orders'.

MALCOLM NEWSON

Newcastle upon Tyne
November 1992

The following individuals and publishers are acknowledged for allowing publication of original illustrations:

T. C. Atkinson: Fig. 3.15(a)
J. C. Bathurst: Fig. 2.8(c)
P. J. Boon: Fig. 7.4(b)
British Ecological Society: Fig. 7.9(c)
A. Brookes: Fig. 5.13
Cambridge University Press: Figs. 5.12, 7.6(b) and 9.14(a)
M. J. Church: Fig. 4.9
J. D. Deans: Fig. 2.6
Earthscan: Fig. 4.1
J. B. Ellis: Fig. 8.3(b)
S. S. D. Foster: Fig. 8.10(b)
J. D. Hewlett/Georgia University Press: Figs. 5.1 and 8.1
N. T. H. Holmes: Fig. 7.8
Hydro Research and development: Fig. 5.9
Institute of Hydrology: Figs. 2.7, 2.9, 2.13, 2.17(b), 3.1, 4.7, 5.3, 9.2(a), 9.7
I. Jackson/Longman Group UK: Fig 3.10(b)
G. A. McBean: Fig. 1.8
P. E. O'Connell: Fig. 9.2(b)
G. E. Petts/John Wiley UK: Figs. 5.14(a), 5.14(b)
S. Trudgill: Fig. 8.4(a)
R. C. Ward Fig. 1.2
D. R. Weyman: Fig. 3.13
John Wiley and Sons, New York: Figs. 2.18 and 4.9

Preface

The Earth, we are told, is a suitable habitat for organic life because of its unique atmosphere, of which water vapour constitutes a vital 25 kg per m^2, helping to feed back conditions supportive to life on the planet through controls on climate. The planet should, indeed, be called 'Ocean', since 70 per cent of its surface is water. Each of us depends entirely on water as the solvent and reactant for all the processes we call life; we are, it is said, water animals which have evolved on to land.

These broad views of the role of water and wet environments on Earth have helped move Hydrology, Oceanography, and Climatology up the scientific agenda, both in research and education. The environmental approach has quickly replaced the narrower, traditional ones of physics and engineering which have, for two thousand years or more, dominated mankind's management of water. *Water in the Service of Man* is a book and an attitude with which I grew up, one which extols the heroic manipulations which are possible with this apparently malleable basic resource.

In this century, however, new attitudes have been struck as a result of short-age, pollution, and the need for conservation. Though the planet should be called Ocean, our need is for *fresh* water, distilled by the global heat-engine. Shortages of fresh water, extending sometimes to punishing and fatal droughts (covered extensively by the media) have taught us to value water more realistically. To the mere hardware of water resources: dams, canals, pipes, and cisterns, have been added economic, legal, and social frameworks. These arose from intellectual and cultural traditions which the physicist and engineer at first found entirely baffling! Later chemists and biologists entered the water-management field when industrial and agricultural pollution threatened to further reduce the volumes available, including that volume already dammed (e.g. acidification of reservoir waters in Scotland and Wales). In many ways the biologists have now emerged at the forefront; *Homo sapiens* has finally admitted the broader-scale and longer-term importance of what were formerly known as 'lower forms of life'; they are equally dependent upon water and wetness. There is, as yet, no book entitled *Water in the Service of Water Fleas* but Part II of this volume stems partly from that viewpoint.

Because of the contemporary changes noted above, this book is written by a physical geographer for others of similar catholicity of scientific interests and of similar humanistic concern for natural environments on Earth. Whilst an inter-disciplinary text is an impossible specification, the geographer's constant search for 'space and time in everything' means that the author's professional experi-ence has been one of a range of methodological frameworks; each is represented here. Most are scientific; we have yet to discover an acceptable replacement for data collection, analysis, and prediction. However, we are beginning to admit that we cannot evoke all the answers in this way: key data are still relatively scarce and there is considerable scope for non-digital values: attitudes, preferences, policies, and laws.

Hydrology entered the curriculum of Geography in higher education in the UK in the 1960s: Roy Ward of the University of Hull, John Rodda at Wallingford, and others must be acknowledged as extremely perceptive in rescuing those aspects of hydrology with clear geographical traits (e.g. patterns of evaporation and rainfall) from their engineering domination. Very quickly the framework of the river basin, already used by geomorphologists, became the dominant one for the study of processes in the field which was to characterize physical geography and the emerging environmental sciences so heavily in the 1970s and 1980s. By 1985 we felt confident enough about the magnitude of the geographical contribution to the study of runoff processes that we began to package it more effectively as a disciplinary statement. Progress has been so rapid and educational interest so profound that we have now included large amounts of hydrology in the school National Curriculum in England and Wales (both in Geography and as an interdisciplinary and field-study theme).

Hydrology dominates and pervades this book, but it is not the mathematical hydrology of dam design or of flood-proofing. It is instead 'environmental hydrology', embracing the basic patterns of precipitation and runoff on Earth and the major ways in which the planet's surface, particularly in the humid-temperate zone, and mankind's occupation of the planet can control those patterns. The river basin or catchment unit is stressed, not only because of its geographical appeal and because it is the basic experimental and data-gathering unit, but because it is the fundamental unit for management. Nevertheless, despite constant reference to the basin (and therefore to land as a hydrological factor) there is also a need to delimit and define water bodies and wetlands as habitat features. 'What is a river?' often becomes a rhetorical question to students, yet it is vital to answer it before 'What is a polluted river?', or 'What is an unacceptably damaged river?'

River life and pollution are treated together here. Whilst chemical determinands are essential to monitoring river quality in terms of danger to humans, we are now more willing to accept the utility of the 'miner's canary' in using biological indicators of river, lake, and wetland health or lack of it. Biological monitoring is often cheaper, more reliable, and a more faithful fit to the space–time framework of runoff and pollution than is chemical monitoring. There is a geography of pollution and knowledge of source–pathway–target sequences in river basins is essential for effective and sustainable solutions to problems of river management.

Finally, this book would be incomplete without cataloguing the *utility* of our current knowledge of environmental hydrology. A problem-orientation puts us on a par with the engineer in the long tradition of water management, but only if we accept the professional rigour which intervention in the environment imposes. The holistic approach which is encouraged in the reader must have a practical and responsible outlet. In the UK our most recent of a long series of water reorganizations has ushered in an era of new strengths in regulatory activity and new opportunities for creativity. The 'think globally, act locally' slogan of the late 1980s has a perfect outlet in our geographical attitudes to, and scientific responsibility for, fresh water. This text is intended to be a simple, practical, and applied work for a wide readership wishing to confirm hydrology as a core environmental science.

Contents

Part I: Hydrology

I. 'Think Globally . . .' 3

1.1 Science, systems, and the hydrological cycle 3
1.2 The global water balance and human dependency: quantities 3
1.3 Global storages and fluxes: modulation 5
1.4 Global hydrological change, past, present, and future 8

2. 'Act Locally . . .': Catchment Precipitation, Runoff, Storage, and Loss 10

2.1 Land phase of the hydrological cycle: quantities 10
2.2 The catchment water balance, components, and calculations 10
2.3 History of the water-balance concept and component measurements 12
2.4 Precipitation 13
    2.4.1 Rainfall: point, area processes, and gauging 14
    2.4.2 Snow: accumulation and melt processes 16
    2.4.3 Precipitation around, on, and under vegetation canopies 18
2.5 Streamflow 20
    2.5.1 Channel-flow processes 21
    2.5.2 Spot and continuous-flow measurements 23
2.6 Simple rainfall–runoff relationships 26
2.7 Measuring catchment-moisture storage 27
    2.7.1 Soil water: gravimetric to neutron probe 28
    2.7.2 Groundwater 29
2.8 Controls on evaporation 30
2.9 Data systems in hydrology 33
2.10 Water balances outside the humid-temperate zone 34
    2.10.1 Hydroclimatic zones 36
    2.10.2 Problems of measurement 36
    2.10.3 Characteristic water balances and the impact of change 36

3. The Basic Threads: Runoff Processes 38

3.1 Components of runoff: volumes and timing 38
3.2 'Bottom-up' treatments of the runoff cascade 38
3.3 Groundwater 38
    3.3.1 Porosity and permeability: hold and yield 39

3.3.2 Groundwater-flow patterns	40
3.3.3 Groundwater and river flows	42
3.3.4 Applied groundwater hydrology and environmental problems	42
3.4 Soil moisture	44
3.4.1 Pores, permeability, and pressure	45
3.4.2 The auger-hole method	46
3.4.3 The special case of peats	47
3.4.4 Soil-moisture patterns and flow	47
3.4.5 Applied soil hydrology	48
3.5 The ground surface	51
3.5.1 Roughness: cultivation and vegetation	52
3.5.2 'Smoothness': urban and paved surfaces	53
3.5.3 Surface-flow patterns: sheets, rills and 'springs' — return surface flows	53
3.6 The stream channel	54
3.6.1 Channel classification	55
3.6.2 Channel processes: transport of water and sediment	55
3.6.3 Channels and river valleys	57
3.7 Assembling the runoff cascade: a model	57
3.7.1 Slope and channel linkages	57
3.7.2 Partial contributing areas	58
3.7.3 Dynamic contributing areas	58
3.7.4 Pipes and seeps	59
3.7.5 Scale effects in the river basin	59
3.8 Runoff processes outside the humid-temperate zone	60
3.8.1 Runoff processes in semi-arid areas	62
3.8.2 Runoff processes under tropical forests	62
4. Runoff Extremes	64
4.1 Extremes in principle	64
4.2 Extremes: mankind and other biota	65
4.3 'Threshold' reactions	65
4.4 Floods and the flood hazard	66
4.4.1 Measurement, data problems	68
4.4.2 Data series, abstracting data	69
4.4.3 Probabilistic treatment of peaks	71
4.4.4 Separating probability from risk	72
4.4.5 Hydrograph analysis/synthesis	74
4.4.6 Flood prediction, forecasting, and protection	76
4.4.7 Floods and nature: a conservation view	78
4.5 Drought and drought hazard	79
4.5.1 Time-scales and spatial scales of droughts	80
4.5.2 Defining droughts: precipitation, soil, river	82
4.5.3 Droughts and river flows: biotic effects	82

4.5.4 Reservoirs, droughts, irrigation, and downstream effects 82

4.5.5 Droughts and irrigation: an international view 83

4.5.6 Predicting drought and climatic change 84

4.6 Runoff extremes outside the humid-temperate zone 85

5. Artificial Influences on Runoff Volume and Timing 90

5.1 History of concepts and experimentation 90

5.2 A return to processes: points and extent of intervention 91

5.3 Fundamental effects: volume and timing 91

5.4 Forests and water 94

5.4.1 Elements of forest land use 95

5.4.2 Regional effects 97

5.4.3 Water use by forests 98

5.4.4 Riparian/phreatophyte effects 99

5.4.5 Predicting forest influences on runoff volume 101

5.5 Agriculture and hydrology 101

5.5.1 Cultivation effects 101

5.5.2 Drainage effects 102

5.5.3 Canopy effects 103

5.6 Urbanization 105

5.6.1 Elements of urban surfaces 105

5.6.2 Urban drainage 105

5.6.3 Experimental evidence at various scales 107

5.6.4 Effects on urban river channels 109

5.7 Land-use hydrology: what next? 109

5.7.1 International practice 109

5.7.2 River-basin land-use data 110

5.8 Effects deriving from water management 110

5.8.1 Water abstraction and use 110

5.8.2 Irrigation 111

5.8.3 Land drainage/flood protection and 'channelization' 112

5.8.4 Reservoir operation: direct regulation of river flow 113

Part II: The Freshwater Environment

6. Link Section: Elements of Freshwater Habitat 119

7. River and Wetland Systems 122

7.1 What is a river? A lake? A wetland? 122

7.2 Quantifying the extent of, and damage to rivers, lakes, and wetlands 122

7.3 Physical, chemical and biological elements of habitat 126

7.3.1 Downstream-zoning of rivers 126

7.3.2 Downward-zoning of lakes	128
7.4 Individual habitat elements and their effects on biota	130
7.5 Biotic indicators of 'naturalness'	132
7.5.1 Plant indicators in rivers	133
7.5.2 Invertebrate indicators	135
7.5.3 Fisheries	135
7.6 Stressed river and lake environments	138
7.6.1 Pollution	139
7.6.2 River regulation	140
7.6.3 Channelization	141
7.7 Wetland systems: Saturated and inundated habitats	144
7.7.1 Wetland classifications	144
7.7.2 Wetland substrates and hydrological processes	146
7.7.3 Vegetative feedbacks: Evaporative loss and interception	148
7.8 Stressed wetland environments and protection	149
7.8.1 Drainage	150
7.9 Material benefits of the freshwater environment	153
7.10 Tropical/semi-arid wetlands and development threats	154

8. Water Pollution

8. Water Pollution	155
8.1 Definitions: source, pathway, target	155
8.1.1 What is 'pure' water?	155
8.1.2 Contaminants, pollutants, organic threats	156
8.1.3 History of water pollution controls	157
8.2 Toxicology	161
8.2.1 Principles and methods of toxicology	161
8.3 Pollution patterns and controls	162
8.3.1 Point versus diffuse pollution	162
8.3.2 Monitoring freshwater pollution	165
8.3.3 Pollution control: contrasts in international practice	166
8.3.4 Towards emission standards? Contrasts in international regulation	168
8.4 Examples of point pollution of freshwater systems	169
8.4.1 Human cycling: water purification and sewage treatment	170
8.4.2 Industrial wastes	173
8.4.3 Animal feeds and wastes: silage and slurry	173
8.4.4 Land disposal of wastes: hydrological pathways	175
8.5 Pollution from diffuse sources	176
8.5.1 Agriculture and forestry as sources: land use versus land management	176
8.5.2 Nitrate pollution	176
8.5.3 Acidification of surface waters	179
8.6 Pollution control in the interests of conservation and recreation	181
8.7 Water pollution and development in tropical and arid zones	182

9. Models and Management

184

9.1 What are predictive models?   184
    9.1.1 A 'model of models' in hydrology   185
    9.1.2 Two models compared   186
    9.1.3 How good are the models?   188
9.2 Principles of river management   189
    9.2.1 River-basin management and environmental protection   189
    9.2.2 Conservation, enhancement, and restoration of the wet
         environment   190
9.3 The context of management: science, people, and planning   191
9.4 Hydrology, environmentalism and global futures   193
9.5 Hydrology and climate change   196

References   199
Index   217

Part I: Hydrology

1. 'Think Globally . . .'

In this chapter we explore the fundamental basis of scientific hydrology: the *global cycle* of precipitation and evaporation which plays vital roles in both the circulation of moisture and the differentiation of the world's climates. We need to know, as far as is possible from available databases, the volumes of moisture in motion and in store at each point of the cycle; first, however, it is important to know the scientific background to the hydrological cycle.

1.1 Science, systems, and the hydrological cycle

Late in the twentieth century it is axiomatic to scientists to assume that there exists pattern and order in circulations of matter and energy. New theories such as chaos now challenge this orderly concept, but for much of the middle of the century the general systems theory of von Bertalanffy (1956) has been a unifying schema for environmental science. The parallel development of the power of computing sustained systematic approaches to instrumentation, data gathering and analysis; the powerful concept of *ecosystem* also became central to the applied sciences of the environment.

Historians of hydrology as a science, such as Nace (1974) point to the many misconceptions about, for example, the source of streams, which persisted long after they were first managed for irrigation and flood protection in the great 'hydraulic civilizations' of prehistoric Mesopotamia, Egypt, the Indus, and the Hwang-Ho. It was not until Perrault's work in 1674 that a coherent cyclical view was taken in hydrology; our subject is, therefore, just over 300 years old! The discrepancy between the youth of the science and the antiquity of its application can be interpreted by observing that manipulation of water is possible with simple applied mathematics and an heroic inventiveness. Thus the need for a coherent world-view of water cycles only became necessary through the dire need for a treatment of resource and hazard problems—

these require prediction or forecasting and for the want of such insights the hydraulic civilizations (heroic manipulators of water) met their demise in environmental stress and civil disorder.

Whilst Pierre Perrault and his countryman Edme Mariotte are credited with putting forward the 'correct' view of a hydrological cycle, the essential empirical studies and observations were conducted by Englishmen: Edmond Halley and John Dalton. Dalton (1766–1844) computed the first regional water balance (of England and Wales) by bringing together the existing rainfall data (which he averaged at 36 inches per year) and an extrapolation of the annual flow of the Thames (13 inches). He went on, less successfully, to measure and estimate evaporation and to research its causes, thus prefacing the scope and emphasis of our contemporary subject.

1.2 The global water balance and human dependency: quantities

We have already noted the appropriateness of the name 'Ocean' for our planet. The domination of Earth by saline waters is spectacular, making the global recycling/distillation 'plant' which deposits freshwater on the smaller land masses absolutely essential to human occupation. Fig. 1.1(a) shows how tiny are the quantities of available fresh water on which we depend, in relation even to ice and snow volumes. It is particularly noteworthy, given the obvious importance of *storage* of a precious resource that the volume of fresh water in man-made lakes now exceeds that in the rivers of the world; note also the relative volumes of water in atmosphere and oceans.

The biological use of water on Earth is sustained by a relatively tiny volume of fresh water and it is therefore of much biological importance that this volume is continuously recycled (Fig. 1.2). The fluxes of water in this cycling process are impressive —see Table 1.1.

Fig. 1.1. (a) The world's freshwater resources as a proportion of planetary water. (b) Geographical detail of water on the planet's land mass. (After Keller, 1984.)

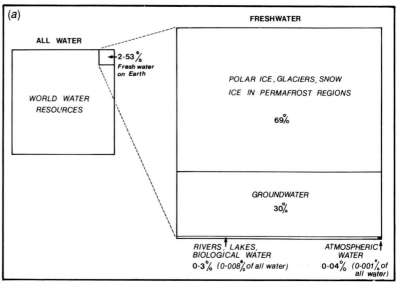

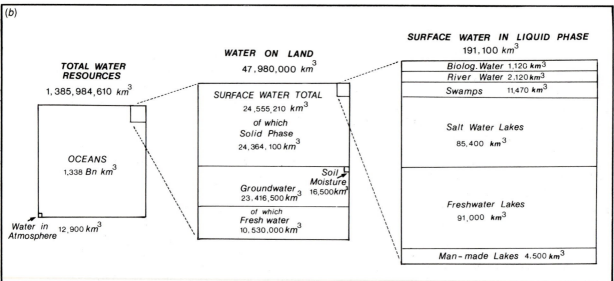

Table 1.1. Fluxes of water (*global hourly averages*)

Precipitation	1–100 kg m⁻²
Infiltration	50 kg m⁻²
Transpiration	1 kg m⁻²
Baseflow in streams	0.01 kg m⁻²

Note: For comparison global biological productivity = 0.1 g m⁻².

It is instructive to consider the nature of human dependency on these fluxes. Human use of water is critical to body function (Table 1.2), to health (working), and to production systems.

Not surprisingly, therefore, 77 per cent of the world's urban population have an organized supply of pure water (1985 data); only 36 per cent of rural people have the same standards. In the case of water for sanitation the equivalent figures are 60 and 16 per cent respectively. Urban water supply is provided at daily rates between 50 l per head (poor squatter areas) and 400 l per head (high-quality detached housing). Industrial production is highly dependent on water for raw material, power, and processing (Fig. 1.3); agriculture is also a major consumer. Direct human modification of the global hydrological cycle is therefore both considerable and extensive.

Fig. 1.2. The global hydrological cycle, illustrating storages (boxed) and fluxes in km³ (after Ward, 1967). Please note that Ward's totals are not comparable with those in Fig. 1.1.

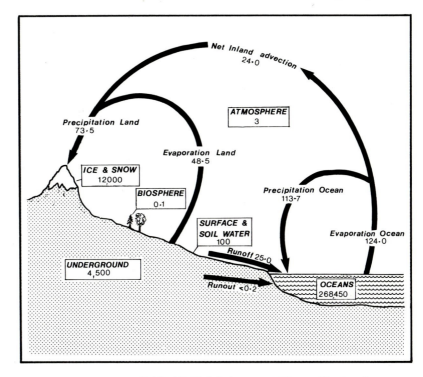

Table 1.2. Human use of water

Storages	
Body:	66% water
Brain:	85% water
Fluxes	
50% drunk as liquid	15% exhaled as vapour
40% from solid food	25% perspiration
10% from body processes	60% urine and faeces

Note: Loss of 10% of body-water renders body immobile. Loss of 20% of body-water causes death.

1.3 Global storages and fluxes: modulation

So far our quantification of the hydrological cycle has been simple and relatively static. Having described biological (mainly human) dependency on a small fraction of global water it is important to consider more details of the way in which the cycle sustains this resource via *storages* and *fluxes* (flows, movements). Table 1.3 lists the major forms of storage (whose capacity is already shown in Fig. 1.1(b)), but it is also important to consider how movement occurs *between stores* and the *residence time* of an average molecule of water in each. The atmosphere has the most rapid turnover of water, followed by rivers. This rapidity not only maintains the natural distribution system for water but results

Table 1.3. Global storages and fluxes of fresh water

Storages	Fluxes
Atmosphere	Precipitation
Oceans and seas	Runoff slopes and channels
Lakes and reservoirs	Evaporation
Rivers	Horizontal vapour flux
Wetlands	Infiltration
Biological water	Percolation
Soil water	Groundwater flows
Groundwater	
Ice	

Residence times	
Atmosphere	8–10 days
Oceans and seas	4000 years +
Lakes and reservoirs	up to 2 weeks
Rivers	up to 2 weeks
Wetlands	years
Biological water	1 week
Soil water	2 weeks to 1 year
Groundwater	days to thousands of years
Ice	tens to thousands of years

in the well-known terrestrial extremes of flood and drought. It is therefore to the slower forms of circulation, such as in lakes and groundwater, that life on Earth looks for reliable and regular supplies. Residence times also have great relevance to the impacts of *pollution*. Damage to lake and groundwater stores is much longer lasting than that to rivers.

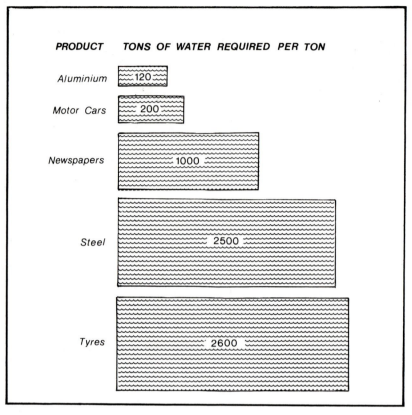

Fig. 1.3. Water in industry: this is often not a consumptive use but may introduce contaminants as it is returned to natural flows.

Self-purification depends on a complete natural turnover in water volume (see Table 1.3).

Progress towards a fully quantitative *global water balance* has been painfully slow since Dalton's measurements (on only a tiny fragment of the global surface): this despite the vital need to establish it in order to fully understand climatic patterns (including patterns of cloud cover which may have a major effect on the progress of the enhanced 'greenhouse effect'—see Fig. 1.4). During the International Hydrological Decade (1965–74) and its succeeding Programme, a co-ordinated international effort at gauging precipitation and runoff allowed refinements to our estimates of global averages for both elements of the cycle. L'vovitch (1973) estimates that global coverage amounts to 50 000–60 000 rain-gauges and 20 000 stream-flow gauges. However, distributions are very uneven and the eventual outcome of global quantifications depends to a great extent on the chosen techniques for interpolating and extrapolating the basic data.

The most recent concise estimate of magnitude for the major components of the global hydrological cycle is by Keller (1974), tabulated here (Table 1.4).

Much of our freshwater resource—almost 70 per cent—is locked up in polar ice and in glaciers. The modulation provided by this storage is of negligible direct importance for human water supply but has great climatic significance. Groundwater's 30 per cent contribution is extremely important; this is the Earth's major natural reservoir exploitable for its resources, and approaches to its protection from over-exploitation and pollution are now entering water-resource policy in most nations of the world. Natural lakes are also important storages, dominating the surface waters of continental areas; they, too, are very vulnerable to damage from abstraction of water supplies, diversion of inflows, and pollution (e.g. Lake Baikal, the Aral Sea, and the Great Lakes).

Most affected of all storages by human intervention and damage are wetlands, rivers, and biological water. Manipulation of these most available resources cannot alter the total volume of water in the global system but by intervening we can profoundly disturb the flux between storages in both space and time. Schemes of afforestation and deforestation alter the evaporation map of a region;

Fig. 1.4. Changes in precipitation over Europe under the enhanced greenhouse effect (assuming that recorded correlations between temperature and precipitation can be used for prediction) (after Lough *et al.*, 1983).

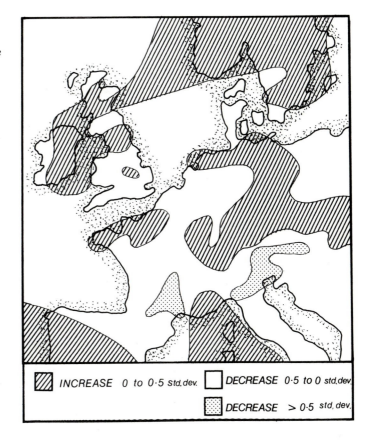

INCREASE 0 to 0·5 std.dev.   DECREASE 0·5 to 0 std.dev.

DECREASE > 0·5 std.dev.

Table 1.4. The world's water resources

Zone of the resources	Catchment area (km²)	Volume of water (km³)	Depth of stratum (m)	% of world's water resources related to:	
				Total resources	Fresh water resources
World's oceans	361 300 000	1 338 000 000	3 700	96.5	—
Land:	148 800 000	47 971 710	322	3.5	
Groundwater (gravitation and capillary water):	134 800 000	23 400 000[a]	174	1.7	
fresh water	134 800 000	10 530 000	78	0.76	30.1
Soil moisture	82 000 000	16 500	0.2	0.0001	0.05
Polar ice, glaciers, snow:	16 232 500	26 064 100	1 483	1.74	68.7
Antarctic	13 980 000	21 600 000	1 545	1.56	61.7
Greenland	1 802 400	2 340 000	1 298	0.17	6.68
Arctic	226 100	83 500	369	0.006	0.24
Mountains	224 000	40 600	181	0.003	0.12
Ice in permafrost	21 000 000	300 000	14	0.022	0.86
Freshwater lakes	1 236 400	91 000	73.6	0.007	0.26
Saltwater lakes	822 300	85 400	103.8	0.006	—
Marshland	2 682 600	11 470	4.28	0.0008	0.03
Watercourses	148 800 000	2 120	0.014	0.0002	0.006
Biological water	510 000 000	1 120	0.002	0.0001	0.003
Water in the atmosphere	510 100 000	12 900	0.025	0.001	0.04

[a] This does not include groundwater resources in the Antarctic which are estimated at 2 million km³ of fresh water.

Source: Keller (1984: 8).

wetland drainage or river-flood protection, or impoundment alter the seasonal and short-term nature of storage.

1.4 Global hydrological change: past, present, and future

Keller (1984) makes the stimulating observation that the atmospheric volume of water at any time is sufficient to meet the annual rainfall of the Amazon region, leaving 98.7 per cent of the planet's surface without precipitation! To keep a cycle going and to spread the input geographically (world average annual rainfall 1030 mm) a total turnover occurs every nine days, a process requiring the energy of large proportions of the solar radiation received by Earth.

The extreme sensitivity of our global atmospheric system in the way it routes energy is now acknowledged; the effects of 'greenhouse' gases on climate have been observed as well as predicted. Much of the scientific evidence for future change is based upon past changes identified from landforms, sediments, or organic remains. Past climate changes had proven impacts upon the hydrological cycle, both total magnitude and geographical patterns. Our principal evidence for change during the habitation of Earth by *Homo sapiens* comes from shifts in the world's rainfall belts during the glaciations of the last one and a half million years.

In the British Isles Thom and Ledger (1976) have produced an approximate and reconstructed water balance for the major periods of the post-glacial (Table 1.5).

Turning to predictions of future climatic change

Table 1.5. Average rainfall and runoff estimates for England and Wales over the last 8500 years

Epoch	Years BP	Rainfall (%)*	Runoff (%)*
Boreal	8500	92–95	86–91
Atlantic	6000	110–115	111–122
Sub-Boreal	4000	100–105	98–108
Sub-Atlantic	2900–2450	103–105	106–110
Optimum	850–700	103	102
Little Ice Age	640–300	93	89
Warm decades	100–50	99	98

* of 1916–50 average.

Source: Thom and Ledger (1976).

resulting from artificial impacts on the atmosphere we are extraordinarily ill-equipped to predict directly the hydrological consequences. As Fig. 1.4 shows, national or continental predictions can be made by comparing warm and cool periods of the immediate past. Global atmospheric models are being constructed for this purpose too but they are very poorly calibrated for the main components of the hydrological cycle:

Our present quantitative knowledge of the global and regional water and energy cycles is very poor. For example, it has not been possible so far to measure either precipitation or evaporation or . . . soil moisture on continental scales. Acquisition of these vital climatological data bases is needed for further progress on global weather and climate predictions and as a necessary step in the study of global change. (McBean, 1989)

McBean's illustration of the global water balance (Fig. 1.5) deliberately draws out the controlling role of the *oceans* and *ice sheets* on the gross circulation of water. Thus, whilst mankind has focused upon, and has profoundly altered the quantity and quality of a tiny fraction of world water (the freshwater

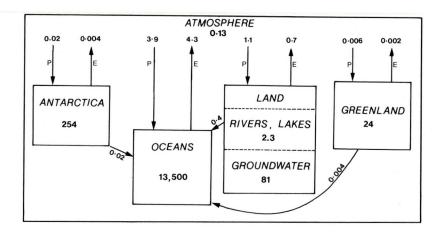

Fig. 1.5. The role of the Antarctic and Greenland ice sheets in controlling the global hydrological cycle during rapid climate change (after McBean, 1984). Units are $10^{14}\,\mathrm{m}^3$; fluxes are annual.

component) its eventual *distribution* under climate change scenarios is modified by vapour and energy exchanges occurring in other regions and at other times. Little wonder that the World Climatic Research Programme has now extended its interests by initiating the Global Energy and Water Cycle Experiment. One of its aims will be to predict the influence of clouds (an essential component of the hydrological cycle) upon the pattern of global warming by the enhanced 'greenhouse' effect.

Returning to Keller's simpler assessment as a means of predicting the resource requirements of the future, it is estimated that $25\,000$ km^3 of fresh water is available for use with existing technology. Total annual human requirements for water (domestic, agricultural, industrial) are approximately 1000 m^3 per person. At present we each have available eight times this figure; with population growth from 3 billion to 6.5 billion by AD 2000 the margin of safety will be less than half that prevailing today.

This analysis is facile in one important respect: it is the distribution of water resources, in both space and time, which is the essential interest to human society and the source of the stimulus to drain, dam, irrigate, or go to war over water. The real challenge to water-resource exploitation and hazard mitigation is to maintain global equity and a concern for freshwater ecosystems whilst coping with the declining overall resource volume, its contamination or pollution, and the alterations to spatial and temporal distributions brought about by climatic change.

As one indication of the care needed L'vovitch (1979) has attempted a quantification of the influence on the global freshwater demand of *changing land use* (see Table 1.6). Water use is achieved at different efficiencies; for example, rice and sugarcane crops 'consume' much more water than other irrigated crops—their productivity is less per unit volume of water supplied. Irrigation systems and many domestic supply systems leak water, requiring much more to be available than is actually 'used'. Large reservoirs in semi-arid lands 'lose' water by surface evaporation.

Nevertheless, L'vovitch estimates an *increase* in controlled river runoff from groundwater and from reservoirs by AD 2000. He cites increasing agricultural use as the reason for this increase in control; natural (flood) runoff will decrease because of its impoundment in such schemes. Agricultural 'use'

Table 1.6. Changes to the global supply and use of water between 1970 and 2000

	1970 (km^3)	2000 (km^3)
Hydrological cycle		
Precipitation on land	110 300	110 300[a]
'Controlled' runoff	14 000	22 500
Surface (flood) runoff	26 800	20 500
Soil water/groundwater	83 500	89 800
Evaporation	71 500	72 800
Water consumption		
Water supply (including industry)	130 (600)	1 050 (1 500)[b]
Irrigation	2 100 (2 800)	4 000 (3 950)
Hydropower	160 (170)	500 (500)
Fisheries	15 (65)	84 (175)

[a] The analysis by L'vovitch assumes that climatic change will not alter the total annual input.
[b] Figures in brackets indicate the volume *extracted* in order to produce the given supply. The shortfall in irrigation can be supplied by sewage.
Source: L'vovitch (1979).

(which effectively means evaporation and transpiration by plants) will increase global rates of evaporation but the author here neglects the impacts of a reduced flux from felled areas of former tropical rain forest. Once again, the geographical location of the resource in question and how it reacts to climatic and land-use changes will be crucial. Water wars are by no means out of the question in regions of extreme stress (see Falkenmark, 1989). Never has hydrology had such a contribution to make to the human hold on our earthly habitat.

A more recent assessment of the anthropogenic effect on the global (terrestrial) hydrological cycle has been made by L'vovitch and White (1990); describing the 'major alterations in the distribution of water on the face of the Earth from 1687 to 1987' they list irrigation as the most extensive influence upon the distribution of water volumes. Impoundment has profoundly altered temporal distributions of surface flows but currently 'no more urgent problem seems to emerge than the question of how present and prospective management can be orientated to cope in practical ways with the rising volume of polluted waste water from factory and field' (p. 249). Clearly, with the relative volume of 'fresh' water naturally so small, there are severe penalties for allowing the figure to dwindle through abuse; there is no better reason for considering hydrology and the river environment together.

We return to this theme in sections 9.4 and 9.5 (pp. 193 and 196).

2. 'Act Locally . . .': Catchment Precipitation, Runoff, Storage, and Loss

2.1 Land phase of the hydrological cycle: quantities

In Chapter 1 we identified the exceedingly small relative quantities of global water involved with that portion of the hydrological cycle which is of prime and direct significance to mankind. The water about which we define our knowledge in this chapter comprises little more than one thousandth of one percent of the planet's volume of the substance (though it is forty times more important as a proportion of fresh water resources).

Before examining the operation of the water balance in river basins and the measurements we make in calibration of the system it is important to know something of the continental distribution of rainfall and runoff and of the scale of river and aquifer basins. Table 2.1 shows a division of the world's rainfall and runoff by continent. Total runoff is, in fact, enhanced by (relatively) small contributions from melting polar glaciers and from groundwater contributions direct to oceans. These contributions represent a 12 per cent increase to around 44 130 km^3 yr^{-1} (L'vovitch and White, 1990). Fig. 2.1 is a map of global river runoff illustrating the importance of the tropical zone; a similar map of groundwater and soil moisture would also

elevate the tropical contribution—indicating a hydrological sensitivity often ignored in the biological concern over tropical ecosystems (e.g. see World Resources Institute, 1990). Whilst little more can be 'harvested' from the world water balance there are of course multiple opportunities to reduce waste and to control water pollution. Many of those opportunities will arise through sustainable management of the world's largest basins (Newson, 1992).

2.2 The catchment water balance, components, and calculations

The word *catchment* itself tells us something of the problems of making hydrological measurements; the word is mainly used by water supply engineers and describes the often small areas (low order—see Chapter 7) which supply storage reservoirs. In the United States the term *watershed* may also be used; the surface areas of catchments or watersheds used for water-supply reservoirs tend to be less than 100 km^2. This is also the scale used for much of the hydrological research from which we draw our knowledge of accurate water balances, flow extremes, and land-use effects. Clearly the very

Table 2.1. Total continental annual water availability and use (km^3 yr^{-1})

	Precipitation	Evapotran-spiration	Runoff		
			Total river runoff	Surface (flood) runoff	Stable (base) flow
Europe	7 165	4 055	3 110	2 045	1 065
Asia	32 690	19 500	13 190	9 780	3 410
Africa	20 780	16 555	4 225	2 760	1 465
North America	13 910	7 950	5 960	4 220	1 740
South America	29 355	18 975	10 380	6 640	3 740
Australia and Oceania	6 405	4 440	1 965	1 500	465
ALL INHABITED CONTINENTS	110 305	71 475	38 830	26 945	11 885

Source: L'vovitch (1979: 201).

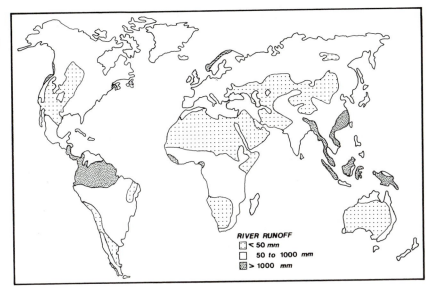

Fig. 2.1. Global distribution of river runoff (after World Resources Institute, 1990).

RIVER RUNOFF
☐ < 50 mm
☐ 50 to 1000 mm
▨ > 1000 mm

large river basins of the world are also gauged but more often than not the purpose of equipping a moderately sized basin of, say, 10 000 km² with both river-flow and rain-gauges is to measure only the available water resource for exploitation, or to monitor that exploitation against the background of natural variability.

Scale, therefore, is a substantive issue in hydrological measurement and as we examine each component of the water balance (below) the reader should consider the practical and analytical differences faced by organizations monitoring and operating on catchments of 10 km², 10 000 km², and so on.

In addition to differences in approach to hydrological measurement wrought by scale, contrasts also result from the wide variation in the *components of the water balance* themselves. For example, snow and ice climates spectacularly constrain measurement of both precipitation inputs and streamflow outputs; at another extreme, semi-arid conditions also polarize the measurement efforts of hydrologists towards the evaporation component—both climates are highly seasonal.

Taking the least extreme regime for catchment calibration—the humid-temperate climate—the following components require measurement (see also Fig. 2.2):

Streamflow
Precipitation
Evaporation
Soil Moisture
Groundwater

Under moderate conditions of climate this list approximates to an order of priorities. *Streamflow* is the essential element of indexing the performance of a river basin and also crucial to ancillary measurements of solid or solute loads (including pollutants). Whilst measuring streamflow without precipitation is superficial hydrology, measuring precipitation alone is not hydrology but climatology. Streamflow and precipitation together, if measured continuously, allow calculation of a basic hydrological index by which river basins may be compared—the runoff coefficient. The coefficient varies in both space and time, mainly between zero and 100 per cent (but storage and 'carry-over' of stored precipitation may produce coefficients of >100 per cent during certain periods). Further details are provided in Section 2.6. Continuous records of streamflow and precipitation also allow the important time parameters of a basin's throughputs to be determined.

Rates of *evaporation* can be calculated by subtraction of streamflow volumes from catchment precipitation; the resulting losses are an integration of infiltration, catchment leakage, measurement error, *and* evaporation but the latter term dominates in most climates. Because evaporation is, in most climates, difficult and expensive to measure, the calculation of losses (i.e. including storage and errors) may prevail, especially in the case of monitoring organizations. By contrast much *hydrological research* focuses upon the measurement of evaporation (see below).

Fig. 2.2. The catchment hydrological cycle displayed as a landscape view and as a series of 'tank' stores which route an input precipitation through to a river flow (shown as time series). An emphasis is put on measurements in the cycle.

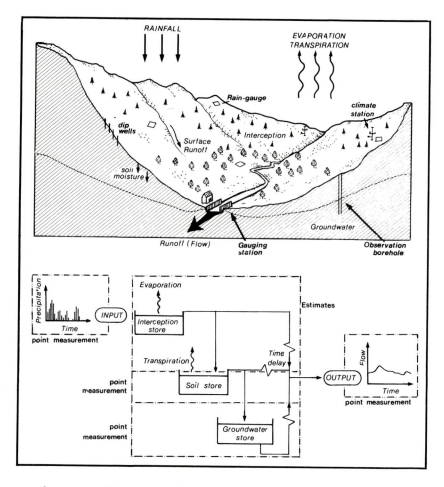

Soil-moisture and groundwater sub-systems of the catchment water balance dominate in some climates, some geologies, and in certain periods (e.g. droughts in the humid-temperate zone). Soil moisture and groundwater have a controlling influence upon runoff *processes* (see Chapter 3); thus, detailed hydrological research, e.g. on land-use effects, requires their measurement. In other circumstances the soil and aquifer rocks beneath serve mainly to introduce the *storage* term to the water-balance calculation. If measurements can be balanced over periods which begin and end with equal storage volume, this component can be ignored, except as a check measurement. For this reason hydrological balances are often drawn up over the period of one flood hydrograph, the seasonal runoff period (e.g. from snowpacks), or the water year between recharge periods (e.g. 1 Oct. to 30 Sept. in UK).

Box 2.1 lists the combinations of hydrological measurements made in the most common calculations summarizing the hydrological performance of

river basins. Further details of rainfall–runoff calculations are provided in Section 2.6.

2.3 History of the water-balance concept and component measurements

In Section 1.1 of Chapter 1 we briefly touched on the relatively short history of hydrology as a science. It was as late as the seventeenth century that the principle of an earth–atmosphere cycling of water was considered as the basis for the supply of streams and springs, yet prehistoric civilizations had measured components, especially river level (e.g. the famous Nilometer gauge at Rhoda, dating back to 3000 BC with usable data from AD 672. John Dalton's tentative water balance for the Thames opened a nineteenth century of significant progress. For example rain-gauge networks in the UK date back to 1729 and evaporation measurements to

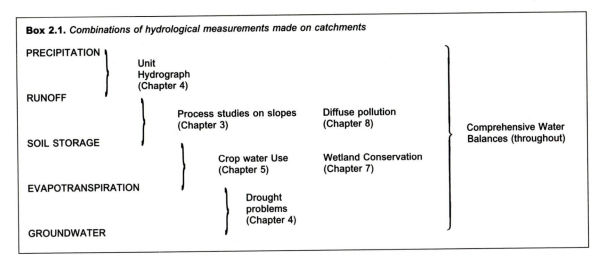

Box 2.1. *Combinations of hydrological measurements made on catchments*

Table 2.2. A brief history of the assessment of water balances in river catchments

Date	Location	Activity
17th C.	Seine basin	Perrault and Mariotte demonstrated rainfall minus evaporation is the source of rivers.
1802	Thames basin	Water balance calculated from measurement of rain, flow, evaporation by John Dalton.
1883	Thames basin	Regular flow measurement at Teddington by Thames Conservancy (British Rainfall Organization had commenced 1860).
1909	Waggon Wheel Gap, Colorado, USA	Catchment experiment to determine hydrological effects of clear felling forest (1919).
1940	Coweeta, North Carolina, USA	First clear fell on multi-catchment forest experiment (continues 1992).
1965	International	International Hydrological Decade (to 1974, followed by International Hydrological Programme). Representative and Experimental catchments set up in many nations.
1980s and 1990s	Global	Use of remote sensing and atmosphere/ocean studies to calibrate global water balance and set context for/checked by river basin studies.

1772, but the nineteenth century saw these co-ordinated and also saw the addition of flow-gauging and groundwater monitoring (spurred by droughts in the 1850s). Whilst fisheries- and pollution-control legislators appealed in 1902 and 1912 for discharge measurements to aid control, the expansion into multi-functional hydrometric networks did not begin until the twin stimulations of water-resource needs (Water Resources Act 1963) and the International Decade (1965–74). The Decade also provided a spur to *research catchments*, particularly with respect to the influence of land use. Kittredge (1973) points out that the interest of hydrologists was first captured by controversy over forests; forest effects on flow and erosion had been observed

throughout the nineteenth century and collected by writers such as Marsh (1864). Finally the Americans began their series of famous catchment experiments at Waggon Wheel Gap, Colorado, in 1909. Table 2.2 very briefly summarizes the historical sequence of scientific, quantitative studies of the water balance of catchment areas.

2.4 Precipitation

Part of a training in hydrology is learning to use the word precipitation (rather than rainfall) to describe the inputs to the ground phase of the cycle.

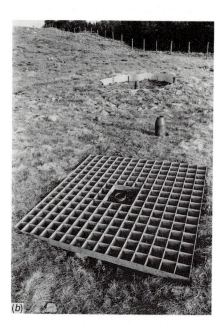

Plate 2.1. (a) Interior workings of tipping-bucket recording rain-gauge. Finger indicates reed-switch operated by small magnet between buckets as they tip. Also shown is the inner edge of a ground-level rain-gauge grid. (b) Site comparing rainfall catches by ground-level rain-gauge (foreground), UK standard gauge (middle), and 'turf-wall' sheltered gauge (background).

According to location the precipitation input may be dominated entirely or partly by rainfall, or by snowfall; in certain special environments the main form of precipitation may be directly condensed water vapour (fog-drip or occult precipitation on to vegetation). To surface-water hydrologists the hydroclimatology of precipitation may have a considerable bearing on the flood hazard, certain airstreams and synoptic circulations defining specific flood conditions. Water quality and ecological research may be profoundly influenced by sources of precipitation, e.g. the sea-salt events produced in coastal regions when rainfall combines with strong onshore winds.

2.4.1 Rainfall: point, area processes, and gauging

In making accurate and repeatable measurements of any environmental variable it is essential to match a knowledge of the natural processes controlling the phenomenon in question and of those modifications which are inevitably made by the instrument and/or observer. In the case of rainfall-gauging we need to know how rainfall originates in rainfall systems (to design gauge networks), how it impacts on the ground or on vegetation canopies (to design individual gauges), and how it enters the rain-gauging device. These needs add considerable complexity to the almost bizarre simplicity of most of the world's rain-gauges which consist of a funnel mounted over a collecting bottle. These *storage gauges* are emptied daily, weekly, or monthly and a conversion made between the volume of catch in the bottle and the equivalent depth over the area sampled by the gauge. As Table 2.3 shows, rain-gauges and their networks are designed with a major purpose in mind and there are considerable differences between gauging for long-term water

Table 2.3. UK rain-gauge spacing to meet data requirements

Data use	Time period	Area (km^2)	Gauge spacing (km)
Water balance	Month	100	7.5–9
Soil moisture deficit	Day	400	3.5–4
Flood design	Hour/minute	1	0.9–1.2
Land-use research	Month	18	0.5

balancing purposes and for, say, flood warning. For time intervals of less than 24 hours simple storage is not appropriate and some form of *recording rain-gauge* is required. Using simple tipping containers, or siphoning mechanisms, increments of rainfall entering the gauge from the funnel are discarded through its base after recording as, for example, a series of 0.5 mm tips, or after making a mark on a paper chart. The so-called *tipping-bucket rain-gauge* is now a major feature of those networks of hydrometric instruments used on-line with computer-based flood-warning systems (Plate 2.1(a)).

Returning to the necessity of matching gauge to process, the point is well made by considering the use of storage gauges in varied topography such as that of the UK. The recommended siting criteria for the UK Meteorological Office's *standard rain-gauge* stress the need to avoid both over-exposure and sheltering. Their formula suggests that any object close to the gauge should be at least four times its height away to avoid the effects of shelter on the gauge catch.

However, over much of the UK's urban and upland landscape it is not possible to meet these criteria, nor can they be met over vegetation canopies taller than the 30 cm by which the UK standard gauge protrudes above ground surface (see 2.4.3). For exposed ground surfaces extensive experimentation in both the field and in wind-tunnels yielded a design of ground-level rain-gauge (occasionally called 'pit gauge') which minimizes the turbulence experienced by the rain-bearing airstream around the collecting funnel. Networks of comparative pairs (standard and ground-level gauges—see Plate 2.1(b)) around the UK have established that in exposed sites the proportion of 'true' precipitation lost from the standard gauge where it is necessarily over-exposed can approach 30 per cent on an annual basis; Rodda and Smith (1986) have mapped the variation of this problem in relation to wind speed. The ground-level gauge design does have disadvantages such as drainage of its pit and performance in snow (see Section 2.4.2 and Plate 2.2(a)).

Table 2.3 indicates, as a control in rain-gauge spacing, the purpose intended for the data. An insight is also required into the importance of local precipitation processes in defining the scale of rainfall inputs to a catchment system. Two basic atmospheric processes yield rainfall in humid areas—*convection* and *frontal activity*; the latter may be increased by *orographic enhancement* where condensation is encouraged by passage over mountains and uplands promoting a subset of growth and

scouring processes within the air mass (see Browning and Hill, 1981). Because convection processes may be highly localized, e.g. over urban heat-islands, rain-gauges to sample high-intensity storms need to be closely spaced in order to increase the probability of recording this 'spotty' phenomenon (convectional rainfall is 'spotty' in space and time).

However, frontal rainfall tends to be more uniform and, if the local variation attributable to orographic effects over high topography is anticipated in design, relatively sparse rain-gauging, with an emphasis on storage gauges, is adequate.

Whatever the configuration of the chosen rain-gauge network and the balance between storage and recording gauges, data-processing will need to gross up individual records to the scale of the catchment. Techniques abound for this transformation and include:

- simple calculation of the mean catch
- weighting individual catches by the area sampled by the gauge (often called Theissen Polygons (see Wilson, 1983))
- weighting by altitude bands
- fitting statistical surfaces through the point data represented by the individual catches.

In practice the results from these techniques appear to differ very little, but virtually every network is different and opinions become highly subjective on the choice of one technique for particular circumstances (it may be possible to consider data-processing before network design as in the 'domain' principle of locating gauges within each significant combination of slope, aspect and altitude within a catchment (see Kirby *et al.*, 1991)).

Moving away from direct techniques of measuring rainfall on impact with the ground, the most promising 'sideways' approach to rainfall measurement utilized the interference produced by precipitation falling through a radar beam. This effect, first noted as a disadvantage to aircraft location in World War II, can be refined to provide estimates of the rain falling through the atmospheric column. Conventional rain-gauge records are necessary to calibrate the system, especially for mountainous terrain where the ground also interferes. However, for flood warning-systems, the benefits of weather radar exceed costs by more than two to one (British Hydrological Society, 1989). As an example, the 2 km grid display available on a VDU (see Plate 4.3), or as a digital input to a flood forecasting model, is shown in Fig. 2.3 for London during a period of

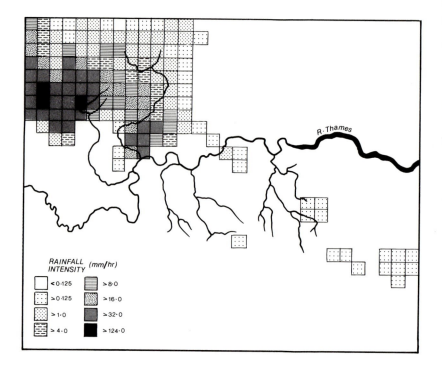

Fig. 2.3. Rainfall intensity as measured by radar at 1735 GMT on 8 May 1988 (after Haggett, 1989).

thunderstorms in 1988. For purposes other than flood warning the direct costs of radar measurements are unlikely to yield similar benefits, but all hydrologists benefit from the improvements to weather forecasts which broadcasters can achieve by their use of weather radar.

2.4.2 Snow: accumulation and melt processes

Fig. 2.4 shows how large a proportion of the populated lands of the northern hemisphere receive a significant contribution by snow to their annual precipitation. However, even more so than with

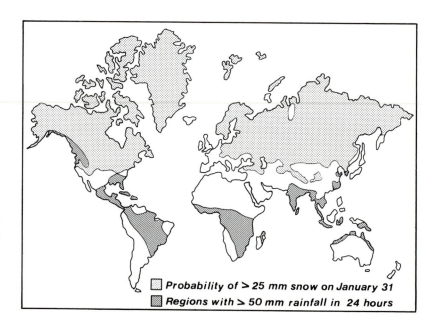

Fig. 2.4. Deep snow and intensive rainfall as sources of river runoff: a global depiction.

Table 2.4(a). Snow-measurement techniques in hydrology

Techniques	Comments
SNOWFALL and DISTRIBUTION	
Rain-gauges	Providing they are not buried
Met. observations	To keep a 'snow diary'
Aerial photography	Attitudinal extent of fall/drifting
SNOWPACK DEPTH	
Snow stakes	Groups of brightly-coloured stakes observed from distance by binoculars
Manual rods	Labour intensive; access problems
Photography	Good for small number of sites
Remote sensing	Good for deep, regular, seasonal snow
SNOWPACK DENSITY	
Snow pillars	Depth as well via density
Coring	Labour intensive; access problems
MELT-RATES	
Heated rain-gauges	Melt snow as it falls and record natural rates of melt
Melt-gauges	Record and transmit

Table 2.4(b). Snowpack densities

Snow type	Density (kg m^{-3})
Wild snow	0.01–0.03
Ordinary new snow, immediately after falling in still air	0.05–0.065
Settling snow	0.07–0.19
Settled snow	0.2–0.3
Very slightly wind-toughened, immediately after falling	0.063–0.08
Average wind-toughened snow	0.28
Hard wind slab	0.35
New firn snow*	0.4–0.55
Advanced firn snow	0.55–0.65
Thawing firn snow	0.6–0.7

* snow partly consolidated into ice

rain-gauging, the techniques of snow measurement are highly variable, with the rather specific needs of hydrological-data gathering often contrasting with those of climatologists (Table 2.4(a) and (b)). Snow, its seasonal accumulation, and melt are important for hydrologists for the following reasons:

- Snow is deposited on the landscape in different patterns from those of rainfall, with land-use influences playing a part in addition to topography; put at its simplest, snow *drifts*.
- The hydrological significance of snow changes during the lifetime of the unmelted *snowpack*, even if this is measured in days; snow changes its *water equivalent* as it compresses and partially melts, a process known as ageing.
- *Thawing* is a complex process but is highly

significant for flooding; it therefore needs both measurement and prediction
- Snow surfaces have special atmospheric interactions influencing *evaporation* and the collection of aerosols which may be pollutants (e.g. acids).

Unfortunately the conventional rain-gauge is not an accurate snow recorder; it measures that volume of snow which has collected in the funnel but only after melting and most rain-gauges become drifted over during substantial snowfalls unless special designs are used (Plate 2.2(a)). The simplest measurement of snow at-a-point is its *depth*; however, this is spatially very variable due to drifting, the influence of land-use (e.g. forest versus clearing), and structures (fences, roads, etc.). It also varies temporarily during the waxing and waning of a pack. Thus, even if the simplest (portable) measuring-rod or (fixed) snow-stake approach is taken it calls upon large labour resources under difficult conditions of observation and access. For this reason automation has always been sought and two simple principles of weighing and photographing have been employed. The *snow-pillow* technique consists of a receiving-plate resting on a sensitive balance which records both additions to and losses from the snowpack; calibration to the *water equivalent* of the snow is direct because weight is measured. An alternative is to bury a safe radioisotope and sense the attenuation of its radioactivity as snow accumulates; attempts have been made to use the natural gamma radiation of the Earth in this way, sensed from satellites. Photography is useful to identify the distribution of snow in the landscape and with careful calibration stereo photography can be used to measure snow depths on slopes visible from a ground vantage point. The simplest, manual combined depth-and-density approach uses a technique of coring and weighing (Plate 2.2(b)) which yields several hundred measurements in a day's work and a considerable potential to investigate and record spatial variability.

If the rigours of sophistication under adverse conditions force observations back to simple manual modes, the measurement of depth from rods or stakes requires a sampling approach which allows for the variability of both depth and density; snow courses (up and down slopes), or domains (as for rain-gauging) are popular strategies. The important hydrological variable of water equivalent or density can be measured by weighing cores of snow, a simple matter with a 10 cm cover but a lengthy process with 3 m. Densities vary during the life of a snowpack and it is critical in many countries

(a)   (b)

Plate 2.2. (a) Problems of measuring precipitation inputs as snow. Enlarged rain-gauge to project above snowdrifts (foreground), search in progress for ground-level gauge (background). (b) Depth/weight measurement of snow cores to establish water-equivalent of shallow, temporary snow packs.

subject to seasonal or occasional snowfalls to make updated measurements or to use melt-gauges to record what is actually leaving the pack as runoff. Evaporation from snow surfaces can be a very efficient process so not all the water present in the pack will become runoff.

2.4.3 Precipitation around, on, and under vegetation canopies

As Chapter 5 reveals, one of the abiding interests of scientific hydrology during its three centuries has been the influence of natural vegetation and crops upon all aspects of the land phase of the hydrological cycle. Indeed, modern concern extends further, in so far as the moisture exchange properties of the tropical rain-forest canopy are thought to control aspects of world climate.

Fig. 2.5 illustrates the problem of adequately gauging the water balance of the canopy alone; it is shown as a forest canopy but could equally be a shorter crop, under study for its influence on soil erosion or to study its irrigation needs. In both the latter cases the 'precipitation' may be provided by a sprinkler system on a test plot. The main problems of canopy-gauging are:

- Airflow over the canopy sets up patterns of precipitation both around and amongst the crop as a result of *disturbance of the boundary layer.*
- *Practical considerations* include the height and variability of some canopies (thus requiring construction of towers) and the interference caused to measurements by leaf litter or by animals resident in the canopy; the canopy also grows!
- The variability of the canopy itself makes the pattern of net precipitation spatially very variable (see Fig. 2.6). Water drains to ground both by *throughfall* and *stemflow.*
- Many canopies are subject to both *rain and snow* precipitation regimes as natural inputs and may also be irrigated.
- Precipitation samples may well be required for *water quality studies* or biogeochemical research.

Beginning once more with the simplest techniques, we can simply deploy the standard rain-gauging equipment outside the crop, in clearings, just above the canopy, and below it. This may work well outside and in clearings but the exposure conditions above the canopy are unlikely to comply with those for standard gauges and the need to sample the spatial variability below the canopy may

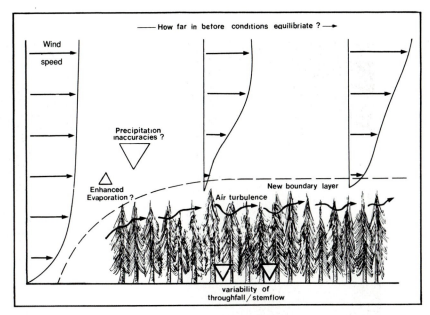

Fig. 2.5. The 'edge effect' of modified airflows and its effects on measurement and processes in forest hydrology.

make the cost of standard gauges excessive. In addition a special approach is required to measure stemflow.

Above the canopy considerable success has been claimed for simple funnel gauges mounted on climbable (for maintenance) towers or pylons (Kirby *et al.*, 1991). However, positioning with respect to the turbulent conditions at canopy level is critical and as such these gauges need to 'grow' with the crop (see Plate 2.3(a)). The catch from such gauges can be led away to a simple storage vessel or into a recording device borrowed from conventional rain-gauging.

In the case of *stemflow*, a wide variety of collars and cups has been designed to fit the trunks of trees and the stems of smaller plants; neoprene rubber is a popular material and within the collar or cup must be a collecting tube leading again to a collector or recording device (Plate 2.3(b)). Growth problems of girth need to be considered and the author has witnessed a number of plants and trees damaged by the constriction of abandoned stemflow gauges!

The high variability of both throughfall and stemflow has prompted innovative designs to integrate sampling across the broad pattern of

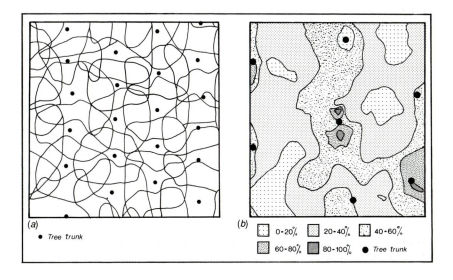

Fig. 2.6. Canopy and throughfall patterns in a conifer forest. (a) Crown projections and their overlap in a 14-year-old spruce forest. (b) Distribution of throughfall around seven of the same trees (after Ford and Deans, 1978).

Plate 2.3. (a) Installation of raised canopy-level rain-gauge necessitated by growth of trees (existing gauge funnel left). (b) Measurement of throughfall (left) and stemflow (right) volumes beneath forest canopy.

increasing throughfall out from the stem (Fig. 2.6). Working on *throughfall* alone, radiating lines of funnels or long troughs have been popular, the former establishing spatial patterns, the latter 'grossing up'. However, some hydrologists have preferred to waterproof the entire sub-canopy space in crops where there is no understorey of vegetation. For example in coniferous forest studies, Calder (1990) and colleagues have operated *plastic-sheet net rain-gauges*, draping the sheets between the trees and attaching them to the stems, with the collected water being led to a tipping-bucket recorder (Fig. 2.7). An advantage of such tall canopies is the ability to gain relatively easy access for instruments and observers; for shorter crops access may be much more difficult and the only option when measuring net rainfall under, for example grass, is to somehow waxproof the soil surface within an otherwise natural plot and lead the resulting flow to a collector.

These grossing-up techniques help solve another problem of net rainfall gauging—that of defining a *representative area* of canopy from which the stemflow or throughfall is deemed to have come in order to assemble a representative figure for large areas of the forest or crop. By including larger numbers of plants in the measured area of, for instance, a plastic-sheet gauge a more representative figure is

obtained.

Two further considerations have influenced the technology of canopy-precipitation measurements—the need to know how much water is stored by interception on the canopy, and whether this interception store is occupied for a spell of time by snow. It has been found possible to deploy gamma-ray techniques in the solution of both problems. The attenuation of gamma rays beamed through the canopy (a measure of its water content) can be related to hydrometeorological measurements made nearby (Fig. 2.8).

2.5 Streamflow

To the uninitiated, streamflow measurement poses fewer problems than do other components of the water balance; apparently all the flow is gathered conveniently in a natural channel whose capacity can be measured, leaving only the velocity to be assessed by field techniques. In reality none of these parameters presents a simple opportunity for accurate or precise measurements. Indeed even the measurement site, chosen for sound experimental reasons or to address an applied problem, can yield problems such as:

Fig. 2.7. An innovative system to use gamma-ray attenuation to index canopy storage of intercepted snow/rain (after Calder, 1990). Below: a detail of the net rain-gauge constructed by draping plastic sheet from rows of trees (collecting throughfall plus stemflow) (after Calder and Rosier, 1976).

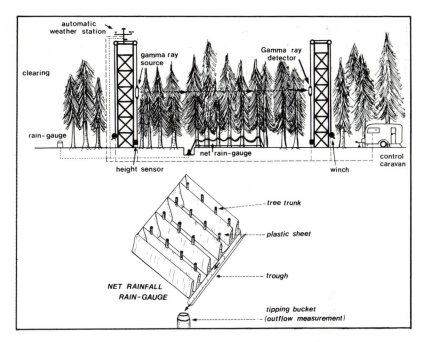

- *seepage or hyporheic flows* beneath and beside the channel;
- *out-of-bank flows* during floods which may also bypass the measuring site;
- poor flow *approach conditions* (see Section 2.5.1);
- artificially affected flows as a result of supply *intakes*, or drain and sewer *outfalls*, or river-regulating reservoirs.

Thereafter an established continuously recording streamflow-gauge is almost certain to be obvious or prominent; it therefore suffers from:

- flood damage
- vandalism
- progressive failure through abrasion, siltation, etc.

2.5.1 Channel-flow processes

To the science of hydraulics, which as we have seen pre-dates hydrology, channel-flow processes represent a largely mathematical challenge. Whilst it is unlikely that hydraulics can perfectly describe flow in the field without recourse to probability (and therefore error) those deterministic relationships which are of well-proven applicability feature largely in the borrowings made by hydrology from hydraulics for the purposes of streamflow measurement.

The most pervasive of these borrowings is the Manning equation which describes the interrelation-ship between gravity in promoting the motion of water in channels and the roughness of the channel boundaries in resisting flow. Discharge is computed from the product of this interrelationship (velocity) and measured channel cross-section, namely:

$$Q = \frac{A \ R^{2/3} \ S^{1/2}}{n}$$

where:

Q = streamflow discharge
A = cross-sectional area
R = hydraulic radius ('A' divided by the length of wetted perimeter of the channel)
S = slope of water surface
n = the Manning resistance factor (chosen from examples in Table 2.5)

In order to see how this equation represents the flow processes which increase or decrease discharge (Q) the reader may wish to raise and lower the value of n whilst holding area, hydraulic radius, and slope constant. This is the equivalent of moving between two similar streams on different sediment types. Alternatively, the investigator may attempt the more difficult test of 'moving downstream' with the equation—this involves increasing area and radius but decreasing slope and n.

The Manning equation is widely used but much criticized (see Hey, 1979). It has good educational value, however, and has practical applications in, for example, the *reconstruction of peak-flood flows* where there are no flow gauges or where these have been damaged or bypassed (see Section 4.4.1). It also provides an indication of the computational task faced by flood-control engineers in designing channels which will cope with (i.e. contain and transmit) flood flows of a tightly specified range.

Problems, however, lie in the choice of a single roughness or friction factor for all flows, even within a single site. In practice the friction exercised by bed and bank materials decreases away from those boundaries and as the body of flow moves further from them during floods then frictional influence also declines. As a result of these factors the typical *velocity profile* in any stream cross-section shows a 'thread' of maximum values towards the centre and towards the surface (Fig.

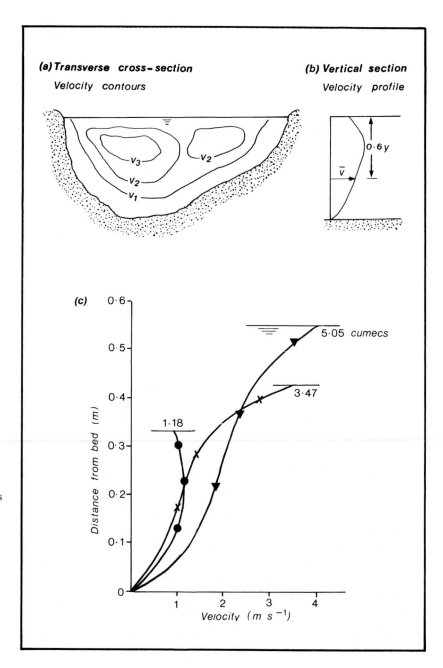

Fig. 2.8. Velocity distributions in a stream cross-section. (a) Transverse section, 'smooth' channel; (b) Vertical section, 'smooth' channel; (c) Vertical profiles in a channel lined by large boulders (after Bathurst, 1988).

Table 2.5. Values of Manning's 'n' channel roughness parameter for various channel materials/ configurations (Canadian National Committee International Hydrological Decade)

Type of channel and description	Minimum	Normal	Maximum
Excavated or Dredged			
Earth, straight and uniform			
Clean, recently completed	0.016	0.018	0.020
Gravel, uniform section, clean	0.022	0.025	0.030
With short grass, few weeds	0.022	0.027	0.033
Earth, winding and sluggish			
No vegetation	0.023	0.025	0.030
Dense weeds or aquatic plants			
in deep channels	0.030	0.035	0.040
Cobble bottom and clean sides	0.030	0.040	0.050
Natural streams			
Minor streams			
Streams on plain:			
Clean, winding, some pools			
and shoals	0.033	0.040	0.045
Sluggish reaches, weedy, deep pools	0.050	0.070	0.080
Mountain streams, no vegetation in channel, banks			
usually steep, trees and brush along banks submerged			
at high stages:			
Bottom—gravels, cobbles and few boulders	0.030	0.040	0.050
Bottom—cobbles with large boulders	0.040	0.050	0.070
Flood plains			
Short grass	0.025	0.030	0.035
Cultivated areas:			
No crop	0.020	0.030	0.040
Mature field crops	0.030	0.040	0.050
Trees:			
Dense willows, summer, straight	0.110	0.150	0.200
Heavy stand of timber, a few down trees, little undergrowth,			
flood stage below branches	0.080	0.100	0.120
Major streams			
Regular section with no			
boulders or brush	0.025	—	0.060
Irregular and rough section	0.035	—	0.100

2.8(a)). Viewed from one side (Fig. 2.8(b)) the velocity profile is bowed, an important phenomenon for locating current meters with depth and for calculating the power of rivers to move their bed sediments. In hydraulically-rough channels such as those with boulder beds, the vertical velocity profile becomes distorted into more of an 'S' shape as flow 'shoots' between the bed sediments (Bathurst, 1988; and Fig. 2.8(c)).

Before leaving the processes which govern channel flow (for further details the reader should refer to good hydraulics texts such as Francis and Minton, 1984), it must be stressed that two extra forms of frictional resistance to flow also influence discharge. The channel may have sinuosity which imparts a form roughness as flow becomes distorted round bends. It should also be remembered that flow requires water to move over water! Thus a resistance is developed by turbulence and by distor-

tions to the flow surface in supercritical flow transitions ('standing waves') and by protruding coarse sediments.

2.5.2 Spot and continuous-flow measurements

A simple yet accurate measurement of the flow rate or discharge of very small streams can be made volumetrically; the equipment necessary includes a tank of known volume and a stop-watch to measure the time taken for the flow to fill it. During droughts, in which even quite large streams dwindle to a trickle and flow remains almost constant for days, volumetric gauging may contribute a very important means of spatial survey and also provide an essential check close to gauging-stations whose normal accuracy is threatened by the very low depth of flow (Plate 2.4).

The more conventional approach is to make

Plate 2.4. Volumetric gauging within an existing weir structure during drought flows.

representative observations of streamflow velocity in a given cross-section with a *current meter* (normally a propeller-driven instrument but electromagnetic meters are becoming more popular). Fig. 2.8(a) shows that care is needed in positioning the meter; $0.6 \times$ depth (measured from the surface) is adopted as a standard depth for average velocity in straight, uniform channels in fine sediments. However, with a bank of current meters it is possible to integrate the variation in the vertical and combine the results rapidly in a field computer. It is equally important to sample the variation laterally and a number of 'verticals' is selected by judgement. In wider rivers the combined needs of sampling vertical and horizontal variations may mean that a full current-metering may take a team of two more than half a day to complete. In wide or deep rivers the meter is slung from a cableway across the river and raised or lowered by pulleys; in smaller streams hand-held meters operated by wading are feasible except in floods. Under dangerous conditions it is valid to use floats from the safety of bridges, timing their progress over a known distance and reducing the computed velocity by 20 per cent to allow for the faster moving surface water.

However velocity is measured, we still require an accurate measurement of channel cross-section to be able to gross up the measurements to discharge. Each vertical produces an average velocity and this

should be multiplied by the cross-sectional area of that vertical, i.e. half the distance to the next vertical on either side (to the bank for lateral verticals).

A marked reduction in measurement effort to gauge instantaneous flows can result from a technique which integrates volume and velocity. *Chemical dilution* offers this facility but is restricted to those streams which produce flow conditions conducive to mixing. The principle is that a chemical or fluorescent dye is introduced in bulk, or continuously at the top of a channel reach in which no flows join the main one and the chemical or dye is continuously sampled at the bottom of the reach. The speed of travel integrates the various velocities of the reach and the sampled concentrations (compared to that known for the input) indicate volume through dilution.

Whether by volumetric, current-meter, or dilution techniques we have only established so far a single, spot or instantaneous gauging of flow. The next variable to sample, with further gauging, is therefore the range of flows typical of the chosen site. We need to establish a calibration curve or rating for the site and the most practical variable with which to link discharge measurements is river level or stage. Subsequently this can be measured continuously with a float-gauge and the level records converted via computations involving the rating curve or its equation (see Fig. 2.9).

There are a number of practical problems associated with the operation of streamflow gauging-stations based upon rated sections of this type. The rating may change progressively over time as a downstream control, for example a shoal of sediment, moves; in the section itself there may be scour or fill of the bed, or erosion of the banks in flood. Consequently, where channel size and construction conditions permit, *gauging structures* are a popular device for network continuous-gauging stations.

The most frequently used gauging structures are *weirs* and *flumes*, although it is possible to construct flow ratings for pipes, culverts, and other artificial flow channels from hydraulic theory or by repeated field measurements. Weirs differ from flumes in that they produce the critical relationship between stage and discharge by obstructing the flow; the gravity head over the crest of the weir can then be related to velocity and, because weirs have a rigid cross-section (either rectangular or triangular), the computation of discharge is simple. Flow conditions in flumes are more complex: a section of critical, high-velocity flow is promoted by a lateral smooth-walled constriction of the flow. Plates 2.5(a) and (b) illus-

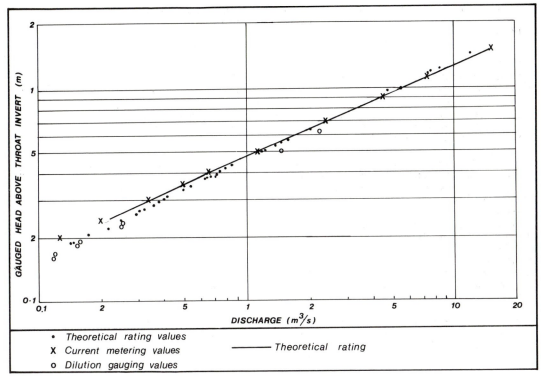

Fig. 2.9. Discharge rating 'curve', against measured water level in a trapezoidal flume (after Smart, 1977).

trate the differences between weirs and flumes; each has its advantages and disadvantages which are brought out, together with those for rated sections in Table 2.6.

An appropriate minor detail of flow-gauging has so far been omitted: that of linking the stream in the measuring section to the equipment measuring water level continuously (normally adjacent to the stream in a protected chamber or hut). Connection is made through a *tapping pipe* whose design is sen-

Plate 2.5. (a) Trapezoidal flume (recorder housed in hut, right). (b) 'V'-notch thin-plate weir, gauging the flow of a forest ditch (recorder housed beneath inverted dustbin).

Table 2.6. Critical conditions of channel and flow for the selection of stream-gauging techniques

Reach/flow conditions	Practical advantages or constraints
VOLUMETRIC Narrow flow-width or pipe; culvert; waterfall to concentrate flow. Drought flows only in larger catchments.	Simple and effective for small flows.
VELOCITY/AREA Uniform flow, straight reach. Avoid boulder or cobble beds where possible. Avoid eroding or depositing sections. Downstream control desirable, e.g. bridge or weir.	Current metering may be time-consuming. Problems in navigable rivers. Encourages frequent visits to station.
WEIRS Interrupts flow—unsuitable for low gradients where water backs up and risks flooding. Unsuitable for sediment-laden flows—merely fills up weir pool. Straight reach, good fall.	Simple construction and simple formulae (H = stage) (B = width) 'V' notch (90°) $Q = 2.5H^{5/2}$. Rectangular $Q = 3.33BH^{3/2}$.
FLUMES Solves the sediment transport problem. Can suffer abrasion damage in sediment-laden flows. Suitable for low gradients. Straight reach.	Formulae more complex and flow conditions can become unsuited to calibration. Sensible to make field calibration.
DILUTION Requires calculation of reach length and conditions to give good mixing. Suited to rough upland channels but not fall-pool reaches.	Getting equipment to site may be a problem or power for continuous measurements.
ALL None suited to very small diffuse flows or very large tidal flows; innovations include electromagnetic velocity measurement and ultrasonic discharge gauges.	All require careful attention to links with the recording system. All subject to vandalism.

sitive and whose maintenance (to avoid blockage) is essential. Many a gauging-station costing tens of thousands of pounds sterling is ruined in operation by poor maintenance. An alternative to a physical link for water levels, growing in popularity, is the emplacement of a *pressure transducer* in the stream section which senses water levels through pressure changes and is connected by cable to the recorder.

2.6 Simple rainfall–runoff relationships

We pause at this stage of the chapter to examine the balance between the two components of the water balance considered so far. As already revealed, not all catchments are fully calibrated for all components of the hydrological cycle (see Fig. 2.2)—in fact most are not, making it necessary to infer a good deal of what we know of catchment processes as a 'black box' system. We know inputs and outputs, hopefully over a common time-base, and to similar levels of accuracy and precision. This time-base should be chosen to reduce the impact on our calculations of variations in catchment storage; we will also want to ensure that 'losses' can be accounted for only by evaporation over this

period—we shall be measuring neither storage nor evaporation. In practice such a period will either be a *runoff season* (or calendar year over much of the humid-temperate zone), or a *flood hydrograph*.

Fig. 2.10 illustrates these two simplistic rainfall–runoff accounting time-scales for a small catchment (for example one draining to a reservoir or used for university research). The figure illustrates a very simple pattern of precipitation over the catchment; three equal-area bands occur between the isohyets for the storm and for the year and the gross rainfall is established by areal averaging.

To compare this depth of rain over the chosen time-scale we convert it to a volume by multiplying it by the catchment area. This calculation shows that precipitation input during the storm which produced the flood hydrograph totalled 175 000 m^3, whilst over the year it was 9.5 million m^3. It then becomes necessary to measure the volume of outflow represented by the flow records for both time-scales. Cumecs (m^3 s^{-1}) are the units of continuous discharge records and so integration beneath the curves will yield runoff totals (remembering to calculate time steps as seconds).

Fig. 2.10 is not scaled for discharge but representative records for the hypothetical catchment in the

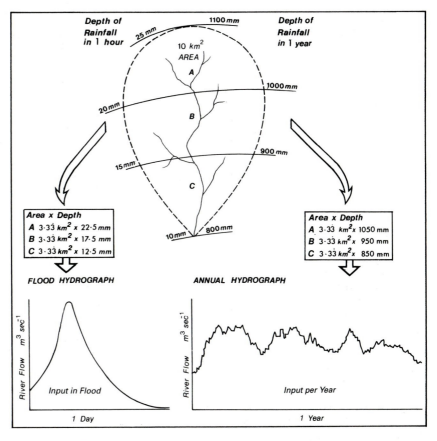

Fig. 2.10. Rainfall runoff comparisons over short, 'event', and annual time-scales indicating the need to convert quantities between depths and volumes for comparisons to be made.

humid-temperate zone might yield runoff totals of 150 000 m^3 and 4.75 million m^3 respectively. These indicate the efficiency of runoff in short storms (about 85 per cent runoff coefficient, or runoff as a percentage of rainfall) whereas over a year evaporation removes half the incoming precipitation (i.e. runoff coefficient is 50 per cent).

2.7 Measuring catchment-moisture storage

Although we have argued earlier that measurements of evaporation are more important to hydrological calibration of catchments a consideration of *storage* is also important for a number of reasons:

(1) One of the major *controls on evaporation* rates is soil moisture, which we therefore consider first for purely educational reasons.

(2) Storage considerations are vital to *water-supply problems*, often under those climatic conditions where evaporation is too high or too low to be of practical management consequence.

(3) Storage of moisture in the cycle is of vital importance to *habitat*; environmental hydrologists concerned, for example, with wetland conservation will concentrate their measurement effort on storage (and perhaps runoff process—see Chapter 3).

There are many forms of measurement of *soil moisture*, depending on whether one wishes to index *content, level of saturation,* or *tension* (Table 2.7). To appreciate the differences one first needs to consider the forms which water adopts in an ideal soil column (Fig. 2.11); having looked at the ideal it is perhaps best to abandon it for practical purposes since, as we observe in Chapter 3, many of the processes of runoff in catchments are set up by the inhomogeneity and horizonation (layering) of field soils.

It is a widespread view amongst pedologists that the planet's soil resources have never received the levels of study and monitoring which befit its most reactive and critical role in terms of human use. Hydrologists would agree; many comprehensive hydrological studies of catchments require specially commissioned surveys or the design of innovative equipment in order to bring soil hydrology to a similar intensity of treatment to that given to

Table 2.7. Guide to the selection of soil-moisture measurement techniques

Soil-moisture variable	Purpose of measurement	Techniques of measurement
Moisture content	Moisture storage Moisture 'accounting' Moisture patterns	Gravimetric: resistance Neutron scattering Indicative use of vegetation. Remote sensing of bare soils

Saturation level (soil-water table)	Depth to water level
	   —drainage experiments
	Depth to water level
	   —runoff prediction
	Patterns of saturation
		Dip-wells
		
		Tensionmeters; piezometers
		
		Indicative use of remote sensing

Moisture tension	Conditions for plant uptake Soil-moisture movement	Porous pot tensionmeters Scanning recording of banks of tensionmeters.

precipitation, runoff, and evaporation. The issue of the global wetland resource and its depletion illustrates the way in which ignorance of basic hydrology has permitted unrealistic forms of exploitation and destruction (see Chapter 7).

2.7.1 Soil water: gravimetric to neutron probe

Moisture content is a basic variable of soil hydrology; it is an important indicator of both the storage term in the hydrological cycle and the available water for vegetation, including crops. It is not an easy variable to measure; there is a considerable conflict between the ease of single measurements and the clear requirement for continuity, best illustrated by the basic gravimetric measurement. Handbooks of soil-survey techniques (e.g. Soil Survey of England and Wales, 1974a and b) describe the details of gravimetric sampling. A few grams of soil are removed in a sealed pack to the laboratory where they are carefully dried after weighing. Reweighing after the moisture is driven off yields the moisture

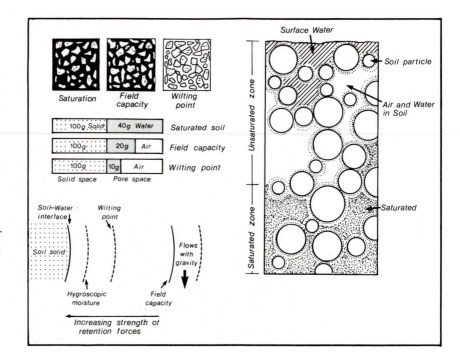

Fig. 2.11. The characteristic states and distributions of soil moisture which have a bearing on measurements of saturation, content, and tension.

content by weight. Further measurement of the bulk density of the soil is required before a volumetric proportion is calculated.

The problems raised by the technique include those of representative sampling (both in the horizontal and vertical dimensions) and the fact that measurement is destructive—it cannot be repeated on the same soil column! For this reason more advanced physical principles, such as the variation of electrical resistance and of neutron scattering from a quantified source, have been deployed in order to give continuous, non-destructive field measurements. There are commercial prizes to be won in this field: irrigation is an expensive use of water resources and farmers making extensive use of irrigation can modify their application rate and timing if provided with on-line soil moisture contents. Of the physical principles so far applied to soil moisture measurement, neutron scattering shows most promise. The principles of a neutron probe device are described by the Institute of Hydrology (1981). If carefully calibrated against both soil and pure-water standards the slowing of the fast neutrons scattered through the soil in a 10 cm horizon from a safe radioactive source can be used to compute moisture contents. The probe is lowered in an access tube through the full soil profile down to bedrock and the moisture contents simply totalled (Plates 2.6(a) and (b)).

Turning to levels of soil saturation, agricultural interest in such measurements has been less than in the case of moisture contents but hydrological interest has grown rapidly in support of process studies such as those described in Section 3.4. The most impervious soil horizon is a saturated one and since saturation has an interesting geographical and temporal pattern (see Sections 3.7.2 and 3.7.3) which controls runoff under certain conditions, the measurement of soil water-tables has become popular. It is also relatively simple and cheap, utilizing *dipwells* driven into the horizon(s) to be sampled. These are commonly 50–100 mm in diameter, lined with perforated pipe, and the water level sensed with a probe from the surface. The probe is commonly electronic, the water surface completing a circuit which makes sound or light. The observer then reads off the depth to the water surface. Such measurements may be used to indicate changes in catchment storage in those catchments (such as those containing deep peat soils) where the soil is either very permeable or very impermeable—in other words where the fluctuating saturation level accounts for most of the stored moisture.

Measurements of infiltration and of soil moisture are more appropriately considered in Chapter 3, having been put into context in Table 2.7. They deal with fluxes of soil moisture (respectively, vertical and horizontal) rather than storage.

2.7.2 Groundwater

Groundwater storage is a phenomenon in all river basins whether or not they are underlain by aquifer rocks. Aquifers (Fig. 2.12) dominate the storage and flow of groundwater; non-water-bearing rocks are known as *aquicludes* and may underlie an *unconfined aquifer* or may *confine* water-bearing rocks in other situations. A widely overlooked group of groundwater storages occurs in shallow aquifers, e.g. glacial drift or alluvium; storage in these may effect a control on the water balance of upland

Plate 2.6.
(a) Measurement of soil-moisture content using neutron probe equipment (probe itself has been lowered into soil here).
(b) Counter/scaler and display of probe readings.

Fig. 2.12. Characteristic zones and processes of the groundwater system.

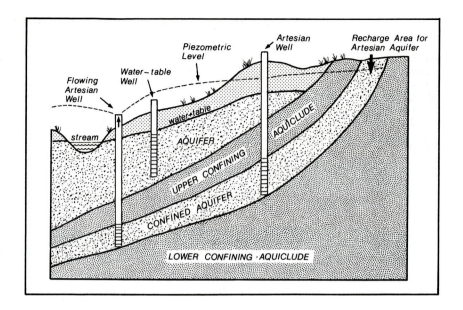

catchments (Hudson, 1988). Another group of poorly understood aquifers are those where fissure flow in joints, faults, or caverns predominates over intergranular flows in the rocks themselves. Karst limestones and some granites are particular examples. The water-table (surface of the saturated strata) may be a fairly regular feature in rocks with the intergranular flows, but in fissured rocks any apparent water-table may be highly localized and an adjacent measurement may be very different in terms of depth-to-water.

In fissured rocks and some shallow aquifers the most suitable monitoring system for changes in storage may involve measuring the flow from *springs*, the points at which the groundwater system yields flow to the surface system. Springs are often valued as a cheap and relatively pure public water supply and so are gauged (using weirs or flumes) for this purpose too.

The most beneficial aquifers for supply tend, however, to be those unconsolidated sands and gravels, or porous sandstones and limestones from which water must be pumped in boreholes (or drawn traditionally from wells). In these cases a thorough network of *observation boreholes* is also established to monitor fluctuations of the water-table, both to assess the resource for abstraction and to plot changes in the storage of the aquifer against precipitation, evaporation, and streamflow. In such cases the results from observation boreholes (see Fig. 2.13) need to be related to gauge records from surface streams to gain a full impression of the

regional water balance. Except in the simplest terrain the groundwater catchment may be different in extent from that for surface flows, a discrepancy which then requires investigation with those techniques which we describe under process studies in Section 3.3, i.e. for measuring and predicting flow directions and volumes.

A final note of complexity is that groundwater systems, whilst generally recharged from rainfall, may be seasonally recharged from *surface streams*. In these circumstances a downstream surface gauge may record less flow than one upstream. In recent years in southern Britain surface streams have been known to dry up completely during drought, a phenomenon blamed on over-exploitation of groundwater leading to lowered water-tables.

In arid and semi-arid climates there is the additional concern that exploitation of groundwater is equivalent to water 'mining'; long-term storage over millennia means that these reserves were recharged under wetter, 'pluvial', conditions and any current depletion is a gross depletion of unrenewable assets.

2.8 Controls on evaporation

This section is so titled because until very recently hydrologists seldom measured evaporation as a flux but instead we measured its controlling conditions, mainly energy supply and atmospheric demand. Evaporation of terrestrial moisture is one of the crucial processes of planetary regulation, playing a part

Fig. 2.13. A characteristic plot of three years' water-level data from an observation well in the Chalk/Upper Greensand, Compton, UK (after Marsh and Lees, 1985). (Heavy line represents current levels, histograms represent statistics of the historical record for the site).

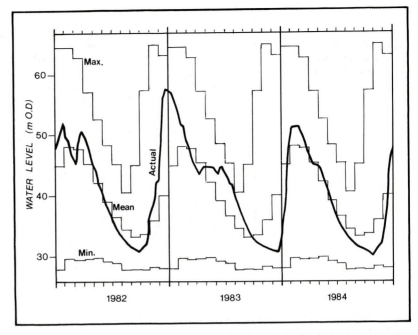

in the global distribution of both energy and moisture. In the terrestrial fresh-water balance, evaporation occurs from three main situations:

- From *open water*—large bodies such as lakes and reservoirs.
- From open-water droplets stored in the upper soil but more importantly *intercepted* on plant canopies.
- From the stomata (openings in the leaves of plants)—this water participating in the physiological process of *transpiration*.

Formerly it was popular in hydrology to write of *evapotranspiration* from vegetation-covered land surfaces, i.e. to bracket the processes of evaporation and transpiration. However, we have begun to be able to separate the two main processes and a recent research focus on interception has revealed the importance of this separation.

The process of evaporation occurs because molecules of a liquid attain sufficient kinetic *energy* to overcome the surface tension of the main body and escape. The vapour exerts a pressure feedback on further escape and so efficient evaporation requires not only an energy source or sources, but a *vapour-pressure deficit* in the air adjacent to the evaporating surface; this deficit is achieved by efficient removal of vapour (e.g. by turbulent airflows). We therefore have, justifiably, two main physical approaches to the measurement (or more, correctly, prediction) of

evaporation: an energy-balance approach and an aerodynamic approach. In tropical and subtropical climates the energy balance, or even its simple representation through air temperatures, may suffice but in temperate and cool climates the aerodynamic approach requires much more attention, especially for tall crops.

The two sets of processes may be illustrated with a simple diagram (Fig. 2.14); as can be seen, the energy balance is complicated and requires not only a partition of radiation, according to the reflectiveness or albedo of the surface in question, but also partitioning of the heat generated by net radiation into that warming the soil and that warming the air.

It is perhaps interesting, in the knowledge of causative processes, to consider how simple, direct methods of measuring evaporation might infringe or disrupt the controlling conditions. For example, a variety of *evaporation pans* has been designed over the years (for which the author prefers the term scientific puddles). These small tanks of water (Plate 2.7) are direct scale models of the lakes or *open-water* bodies whose surface evaporation they seek to represent. However, they are subject, on land, to different radiation budgets if the surrounding land has a lower albedo than water, to conduction of heat through the base and sides, and to air turbulence quite unlike that of a natural water body. In addition open water does not constitute the

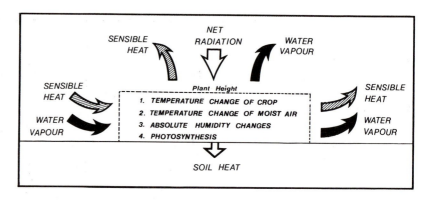

Fig. 2.14. Processes governing the potential for evapotranspiration and indexed in the Penman method for calculating PE$_t$.

evaporative surface for large parts of our continental area where soil conditions or plant moisture processes will control energy and vapour conditions.

To mimic the conditions affecting soil/crop evaporation a number of *evaporimeters* have been designed, the simplest of which is the Piché (Winter, 1974). Leaf (or soil) surfaces are represented by a filter paper from which evaporation occurs, drawing down the water column. Here the environmental 'offence' is to grossly misrepresent the evaporating surface and its resupply from the water store, and additionally to offer a surface which is especially prone to very efficient aerodynamic circulation, drastically increasing rates of loss from the instrument.

Plate 2.7. Evaporation pan showing measuring well (to still surface ripples), thermometers, and anemometer. In the background are two drainage lysimeters.

There are, with both techniques, methods of correcting the records or ensuring that environmental conditions are improved (e.g. floating the pan in a lake, or sheltering the evaporimeter in a plant canopy), but there are also practical problems of maintenance (e.g. birds drinking from pans).

Not surprisingly, therefore, indirect methods of computing the potential evaporation have both a theoretical and practical advantage over the simpler direct measurements. The best-known indirect method is the formula proposed by Penman (Penman, 1948; MAFF, 1967) and successively modified as field-tests and further research became published (Monteith, 1965; Monteith and Unsworth, 1990). Whilst the Penman combination of energy balancing and aerodynamic approaches is potentially greedy for data, the assumptions can be realistically simplified so as to make its calculation possible from 'normal' climate measurements (Table 2.8). Nevertheless, the formulae output potential, not actual evaporation (evapotranspiration or E$_T$), and there are a number of inadequacies in it which put error between potential and actual rates.

The main inadequacies are:

(1) The formula seeks to represent soil/vegetation surfaces (a separate version exists for open water) but does not properly index the restriction of actual rates *when the soil is dry.*

(2) The basic formula does not cope well with the increasing influence of *aerodynamic conditions in tall crops* especially where they store intercepted precipitation.

We address aspects of the second problem in Section 5.4 but the first problem continues to aggravate prediction of actual evaporation. There exists a relationship between the ratio of actual to potential evaporation and the supply of moisture to the evaporating surface. This is clearly not a problem in open water, or in some wetland and

Table 2.8. Measurements required for calculation of potential evapotranspiration by the Penman formula

Measurements of:	Purpose	Instrumentation
Radiation	Quantifying energy available for heating, photosynthesis and evaporation	(Tables of latitudinal radiation) Radiometer Sunshine recorder
Saturated and actual vapour pressures	Quantifying opportunity for vapour to leave water body	(Tables of vapour deficit for temperature). Wet-bulb thermometer
Air temperature	Proportion of energy transferred as sensible heat. Link to vapour pressure	Thermometer (soil temperature may also be measured)
Wind speed	Aerodynamic component turbulent transfer	Anemometer (Estimates of aerodynamic resistance)

phreatophyte plant stands, but in normal land situations the form of the function linking these two indicates a suppression of the actual rate as the moisture store is depleted. Fig. 2.15 suggests that the function varies with soil type; it will certainly vary with the rooting depth of vegetation and hence

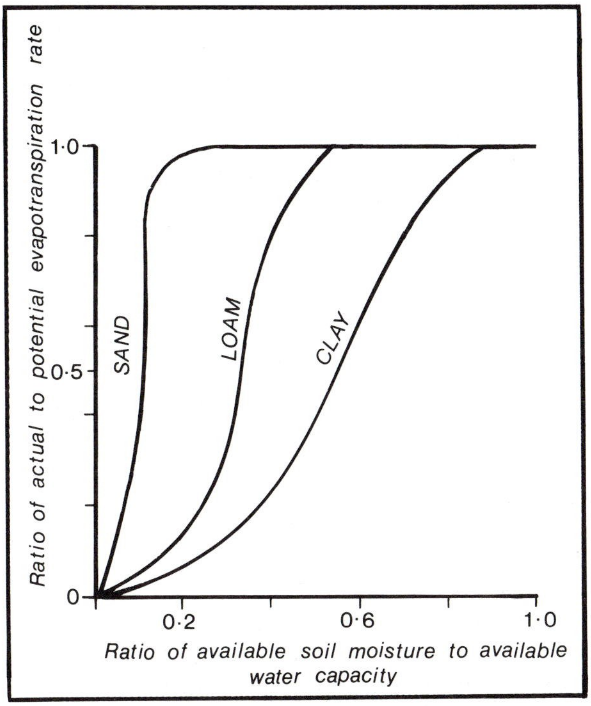

Fig. 2.15. The operation of 'root constant' processes in limiting actual transpiration under drying conditions in three soil types.

the notion of *root constant* has been applied in calculations of soil moisture deficits for use in flood studies and irrigation applications. The root constant determines the deficit at which actual evaporation comes under the control of moisture supply.

Much of what we know of the supply controls on evaporation and transpiration has come from instruments known as *lysimeters*, enclosed areas (columns, tanks, or plots) of soil and vegetation, open to the sky, from which infiltrated moisture can be collected and moisture storage measured. If precipitation is also known the full water balance of a controlled patch of the Earth's surface can be calculated and evaporation measured by the difference between inputs, storage and other outputs. Fig. 2.16 shows three designs of lysimeter; drainage lysimeters are normally the largest and may enclose a hundred square metres or so; Calder (1990) has made very successful use of one in studies of forest evaporation.

Finally, direct measurement of evaporation as a vapour flux, long the ambition of environmental physicists, has of late become a reality, though once more a range of sophisticated information is needed on wind speed, humidity, and temperature to model the flux by momentum transfer.

2.9 Data systems in hydrology

Whether it be the research hydrologist sitting beneath a forest canopy monitoring the hydrometeorology of six levels of air above him or her, or a

Fig. 2.16. Three designs of lysimeter: (a) Drainage type; (b) Pressure weighing type; (c) Balance.

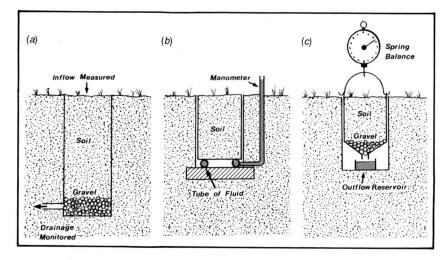

large river-basin authority operating a flood-forecasting model in real time, the science of hydrology is live and multivariate. It is, therefore, impossible to avoid the design of *sensor and recording systems* to make efficient data capture and to permit *analysis* (Fig. 2.17(a)).

An important but much-neglected interface between capture and analysis is *data-quality control*; bearing in mind our frequent reference above to error-prone measurements and the drift in efficiency or calibration of basic instruments it is foolish to neglect either the operation of, or funding for, quality-control procedures. These include:

- Frequent checks on, and maintenance or replacement of field instruments.
- Frequent checks that data are being logged correctly, or transmitted to base.
- 'Pre-analytical' routines of graphical inspection or checks against possible ranges for each variable.
- Analytical tests of all data-streams together, even if full analysis will occur only at the end of the experiment or period of surveillance.

Such systems tend to be most thoroughly developed where national agencies are charged by governments with operating hydrometric networks. The investment made by such agencies and the longer time-scales over which these networks operate tend to force a better discipline of care and maintenance than is possible, or even necessary, during the short life of hydrological field experiments, where funds and labour are often in short supply.

Nevertheless, generalization is dangerous, particularly because of the huge geographical variability in the attitudes of, and resources available to, agencies

running hydrometric networks. For example, much of Africa lacks basic hydrometry and the World Bank sponsored network-design studies as recently as 1990. In North America and Western Europe the reverse has been true: cost-cutting governments have reduced resources to the point where rain-gauge networks have been rationalized (see examples in Shaw, 1989) and flow-gauges taken out (see Fig. 2.17(b)).

As well as long-term archiving of hydrological data, modern hydrometric networks offer an operational use as catchments become managed in real time for such purposes as river regulation, flood control, and pollution control. It must be admitted that the practices of hydrological measurement have benefited from operational use because of the vital importance of fail-safe operation, sophistication of technique and because of the political acceptability of investment; in the UK, for example, river-gauging in support of flood forecasting has become very sophisticated, as has flow-gauging in sewers and polluted outfalls. At last, however, the acceptability of archiving is increasing with the realization that the climate change implies non-stationary (or trending) data records through time.

2.10 Water balances outside the humid-temperate zone

We have concentrated so far upon hydrological measurements made under conditions of abundant, evenly distributed precipitation (falling mainly as rain), relatively low, seasonal evaporation rates, and free-flowing rivers. These were the conditions which

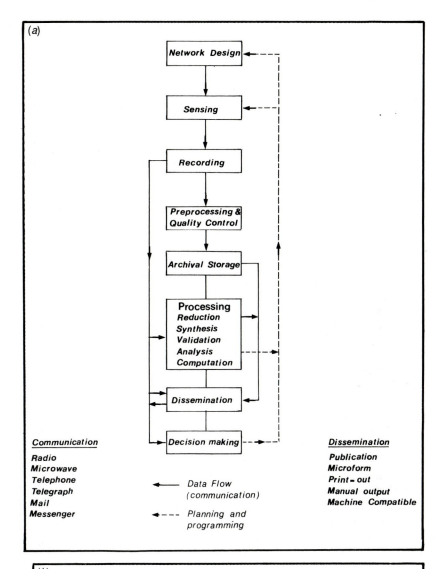

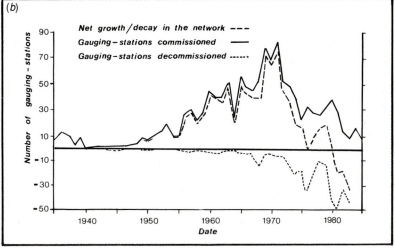

Fig. 2.17. Data networks and processing. (a) The chain of activity between network design and use of hydrological information in management; (b) The rise and fall of the UK river-gauging-station network (after Lees, 1985).

spawned the development of scientific hydrology in Western Europe, but are not those which represent many of the applied-science problems faced by hydrologists today. This section attempts to rebalance the bias of our treatment but is inevitably foreshortened in comparison with the magnitude of the problems it raises.

The main characteristics of the humid temperate zone which facilitate the establishment of water-balance measurements at the catchment scale are:

(a) Rainfall is relatively easily measured because it is well-distributed in space and time (though there are complications from gauge exposure and from temporary snow packs—see Section 2.4). There are few problems in the temperate zone from deep snow packs which bury gauges, high rates of evaporation to 'lose' the rainfall catch from the storage gauge or of heavy, highly localized downpours; all these phenomena are characteristic of other climates.

(b) Rivers are free-flowing, relatively small in scale and accessible for gauging; their range of flows can be mainly accommodated by convenient gauging structures or channel sections.

(c) Evaporation is, however, more problematic in the temperate zone than in others because conditions do not favour the potential rate after soil deficits develop in summer, subtleties of land use and land management may affect rates, and aerodynamic factors are important in an advected climate (see Section 2.8).

2.10.1 Hydroclimatic zones

Jackson (1989) lists six tropical rainfall zones based upon the total input to the hydrological cycle (i.e. wet versus dry) and the seasonality of the fall (single or double rainy seasons or arid). However, he also stresses that *rainfall variability* (<10–40 per cent) between years is of critical importance to agricultural systems; one could add rainfall intensity, important to both soil erosion and river flooding as a further basis for classification.

The Köppen climate classification offers just five precipitation divisions—tropical rainy, dry, mild humid, snowy, and polar. These result from a combination of the basic Hadley Cell circulation of heat, moisture, and momentum in the atmosphere, and the proximity of the zone in question to oceans or mountain barriers, a pattern which demonstrates that there is little local recycling of atmospheric moisture. The exception is the tropical-forest zone, in which almost half the moisture recycles in an average period of 5.5 days (Oyebande, 1988).

2.10.2 Problems of measurement

Remaining with tropical forests, consider the importance and difficulties of making adequate precipitation measurements in a situation of dense forest cover 30 m high, with five canopy layers and an interception rate of 7.5–38 per cent in a region with poor communications and generally low technology.

In the semi-arid climates of the world rainfall is difficult to gauge for other reasons, notably its 'spotty' distribution, its intensity, and the high rates of evaporation which may remove much of the catch from unprotected storage gauges—a film of oil is often added to the rainfall collector. By contrast, evaporation can be measured by evaporation pan because it is the open-water rate which is of most concern and the process is mainly radiatively driven.

Flow-gauging in the episodic, intense floods of the semi-arid zone can be both difficult and dangerous. For example, the Wadi Dhamad (Saudi Arabia), described by Richards *et al.* (1987) is 1000 m wide and carries a flash flood event of up to 2000 cumecs at velocities of 4–5 m s^{-1}. There is little alternative to using the wadi cross section and an estimate of velocity using floats to confirm predictions based on slope and friction factors measured at lower (or zero) flows.

Cold-zone hydrology faces the problem of estimating the precipitation inputs from what may be a substantial pack. Remote sensing has been deployed for many years in providing the critical data on snow cover; image processing can also approximate snow depth, and if density can be gauged by periodic field measurements by helicopter one has the basis for official predictions of the size and timing of the inevitable snowmelt flood. Gauging such a flood can be especially problematic if the river is at first ice-covered since rafts of broken ice are moving downstream at high velocity! In China the air force is employed to bomb ice-gripped rivers to ease the flow.

Further developments in the field of remote sensing and image processing are likely to bring rapid progress to our data bearing upon the gross characteristics of non-temperate climatic zones, though they are unlikely to bring a greater understanding of individual processes.

2.10.3 Characteristic water balances and the impact of change

Strahler and Strahler (1992) depict the Earth's water balance in ten-degree latitudinal steps (Fig.

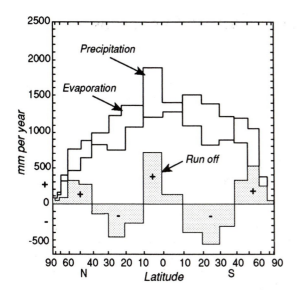

Fig. 2.18. The world water balance in latitudinal zones (after Strahler and Strahler, 1992). Runoff is the product of precipitation and evaporation.

2.18); the graph indicates immediately the importance of the distribution of the atmospheric circulation and of land/ocean bodies. Only two zones, tropical and temperate, show surplus runoff. Rivers flowing from these can supply zones in deficit (e.g. the Nile) and the latter may also benefit from groundwater reserves stored during shifts in the zonal pattern of humidity and aridity, notably during the Pleistocene glacial epochs.

Land-use changes are also of special importance in some non-temperate regions. For example the harvesting of tropical forests (Reynolds and Thompson, 1988) can (but does not invariably) reduce local rainfall (Meyer-Homji, 1988). In snowy climates hydrological impacts of both afforestation and deforestation can be profound. Shiklomanov and Krestovsky (1988) calculate that water equivalents of snow inside forests can be 1.1 to two times those in adjacent fields because the forest is a more efficient depositional environment.

3. The Basic Threads: Runoff Processes

3.1 Components of runoff: volumes and timing

Much of the hydrology described in Chapter 2 treats the catchment as 'lumped', i.e. as a single spatial and temporal entity. We separated component parts in so far as they influenced the bulk performance of the catchment, specifically its water balance; we now move to consideration of a *distributed system of runoff sources*, and the *volume* and *timing* of source-area runoff in contributing to flow in channels. This level of detail, whilst not of direct practical benefit to traditional hydrological applications has considerable importance in environmental hydrology because it aids the prediction of chemical pathways—it also demarcates sensitive zones of catchments.

The simplest depiction of runoff components in the humid-temperate zone catchment is as a series of tanks, stacked vertically, which spill in series down a cascade; some tanks also spill laterally, and from some evaporation is depicted as a vertical loss, the reverse of precipitation (see Fig. 3.1). Clearly the capacity of each tank and the levels at which it spills vertically or laterally determine the volume of runoff from that source in a given flow hydrograph. The timing of runoff from each source is also dependent on the capacity of the tank but is mainly controlled in a real catchment by the spatial arrangement of the tanks and the velocities of flow attained under the characteristic boundary conditions (e.g. intergranular flows in the soil, open-channel flows in concentrated surface flows, or rills). Fig. 3.2 illustrates the characteristic velocities and response times of the main runoff sources during the passage of a minor flood hydrograph in a small humid-temperate catchment.

Clearly if one were to delimit 'geographical hydrology' at a catchment scale, research into catchments as distributed systems would be close to its core and it is not therefore surprising that many hydrologists from a geographical background have helped advance the study of the space–time elements of runoff processes (e.g. Weyman, 1975; Kirkby, 1978; Anderson and Burt, 1985).

3.2 'Bottom-up' treatments of the runoff cascade

Following a building-block approach to what we may call '*runoff domains*' in a catchment it is tempting to deal with surface routes first—they are, after all, likely to provide much of the fastest responding runoff when rain falls or snow melts. However, for the majority of time in most environments rain is not falling (in the UK rain falls for between 4 per cent and 20 per cent of the time) and river flows are sustained by deeper sources. For this reason we make our treatment bottom up.

The relative constancy of the groundwater supply to river systems makes the *baseflow* of rivers a linked property of the solid or drift geology; baseflow reflects the permeability of aquifers in the basin and it takes on their hydrochemical properties. Seen against this relatively constant background runoff source the more superficial sources, stimulated by rainfall events and episodes seem much more ephemeral. Fig. 3.3 illustrates a hypothetical, *annual hydrograph* from a humid climate and demarcates the prevailing influences on the component parts of the river regime.

3.3 Groundwater

Groundwater is crucial to mankind; it constitutes 97 per cent of our global non-frozen fresh water. The constancy of springs has long attracted reverence from human cultures; the purity and coolness of their flow revived travellers and stimulated settlements. Each of these features of the groundwater system: reliability, freedom from impurities and cool, stable temperatures, bespeaks an aspect of the physics of groundwater flow. If we maintain the analogy of a tank in our model, the groundwater tank is very large; where aquifer rocks are present, flow through the tank is slow and may be accessing considerable distances underground. For instance,

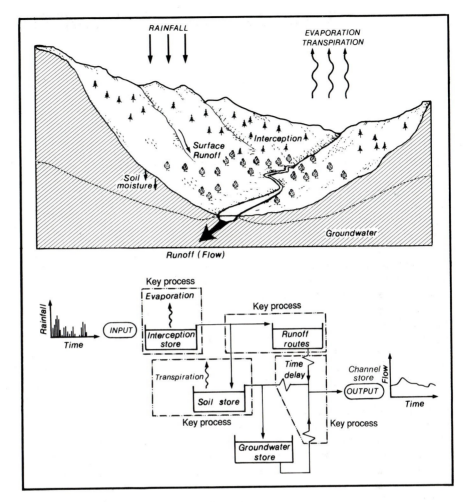

Fig. 3.1. A repeat presentation of the catchment hydrological cycle (see Fig. 2.2) stressing hydrological processes.

recent studies of groundwater pollution by nitrates (the chemical acts as a tracer) has revealed that it may take percolation from 4 to 50 years to penetrate the unsaturated zone down to the water table.

3.3.1 Porosity and permeability: hold and yield

The fundamental property of rocks which makes them function as aquifers, or not, is *porosity* (Price, 1985). At great depth (kilometres) in the Earth's crust there are few if any pores but nearer the surface porosity (expressed as a volume proportion of the rock) varies strongly with rock type (Table 3.1). Pores alone, and their size, are less important than their connectivity—the storage potential offered by a myriad of small 'tanks' must be open to be filled by *percolation* from surface precipitation and to empty to springs, or to wells and boreholes for human supply. This concept of water movement

through aquifer rocks is expressed in their *permeability*. Permeable rocks are able to throughput substantial volumes of water in a relatively short time (nevertheless measured in years in large catchments); they may not do this exclusively through pores and many of the most permeable rocks are not porous. Instead the structure of the rocks, such as the bedding, jointing, faulting and the development of caves in calcareous rocks, boosts the component of *fissure flow*.

Fig. 3.4 illustrates, for unfissured 'pure' materials of various grain sizes, the variability of porosity and *specific yield*, i.e. the flow from an aquifer from lowering its water-table by a measured amount. Put simplistically, specific yield expresses how much one can 'get out of the sponge' when it is drained down on the side of the bath—clearly porosity is not everything and yield reaches a peak where *storativity* also peaks.

3.3.2 Groundwater-flow patterns

In Chapter 2 we briefly investigated the hydraulics of surface flows and linked their velocity to the friction or roughness factors on channel bed and banks and to the gravitational energy available (through slope). The equivalent basic tool of groundwater flows is *Darcy's Law* of 1856. It states that the rate of flow per unit area is proportional to the gradient of the potential head in the direction of flow. The 'proportional to' term is fixed by *k*—the *hydraulic conductivity*, which can be measured in field tests and which varies across several orders of magnitude (Fig. 3.5). So, in calculating the discharge of groundwater (not including fissure flows) one takes not a channel, since there are none, but an imaginary cross-sectional area and multiplies it by the potential head and the hydraulic conductivity:

$$Q = k \, A^{H/L}$$

where:

Q = groundwater flow
k = hydraulic conductivity
A = cross-section of rock
H/L = gradient of potential head.

The chosen cross-section is normally a square metre or the entire cross-section of the chosen aquifer strata draining to a river.

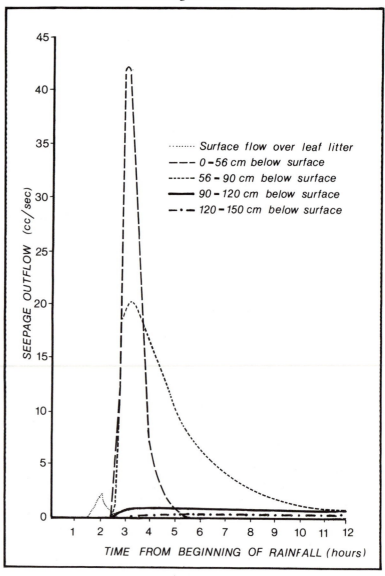

Fig. 3.2. The delays imposed by infiltration and throughflow in a permeable soil with well-defined horizons.

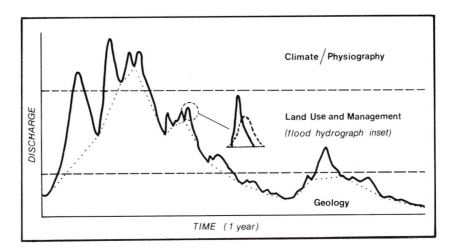

Fig. 3.3. The annual hydrograph of a river (highly simplified) to illustrate the broad controls on volume and timing of runoff. Dotted line indicates an extremely simple separation of the baseflow, whose proportion of the annual flow total constitutes a baseflow index.

Table 3.1. Representative values of porosity for unconsolidated and consolidated natural materials

Material	Porosity[a]
Soils	0.30 – 0.50
Fine sandy alluvium	0.45 – 0.52
Gravel	0.25 – 0.40
Shale rock	0.05 – 0.15
Crystalline limestone	0.01 – 0.10
Chalk and oolitic limestone	0.05 – 0.30
Slate	<0.01 – 0.05
Granite	<0.01 – 0.10

[a] Expressed as decimal fraction volume proportion of rock.

Clearly k, often erroneously called the permeability of a rock (but this includes fissure flow and Darcy's Law applies only to granular media), is unknown in most practical situations. Over the years rock samples have been tested in the laboratory or by pump tests in the field so that k can be tabulated for standard grain sizes (the Law applies to soils also). Fig. 3.6 illustrates tests applied to measure k in both the field and (on rock samples) in the laboratory.

Assuming, for the moment, that Darcy's Law applies to the aquifer in which we are interested (unfortunately this is seldom so for many environ-

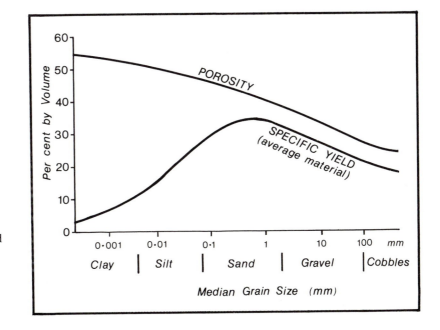

Fig. 3.4. The relationship between porosity and specific yield for unconsolidated sediments in the western United States (after Dunne and Leopold, 1978).

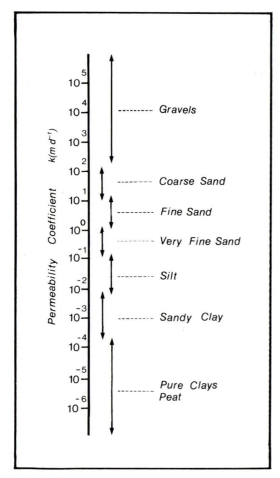

Fig. 3.5. The range of permeability (saturated hydraulic conductivity) in natural soils.

mental applications), a prediction of the direction and rate of flow in any aquifer can be obtained from mapping the elevation of the *potentiometric surface* in the rock. This surface is the *water-table* in unconfined aquifers but in confined aquifers a pressure may build up (artesian pressures are a familiar example) and the water level will rise up an observation well sunk to it. The potentiometric surface is the contour map controlling subsurface flows and upon it can be drawn inflow nets which help predict where flows will occur and how much wells and boreholes will yield.

Another famous Law (Murphy's!) operates in many aquifers to complicate groundwater-flow patterns and hence both water yields and water quality (e.g. where a pollutant has entered the aquifer). This corruption to Darcy's perfect world is intro-

duced by *fissures* (as is the case very frequently in soils—see Section 3.7.4). Flow through fissures in porous rocks is effectively a 'bypass' route giving less opportunity for chemical reactions during the slow passage through pores and less filtration of any solid impurities. In rocks with large fissures, such as the *joints* and *bedding planes* of some sandstones (and certain igneous rocks), and the *cavernous* limestones of the world, any use of the Darcy concept is erroneous and field tests will often employ water tracing with dyes or spores (see Ford and Williams, 1989).

3.3.3 Groundwater and river flows

We have already stressed that groundwater keeps rivers going in periods without precipitation. Only in certain rock types will this *baseflow* enter the river channel from a discrete *spring*. Elsewhere surface streams receive inputs from groundwater by seepage across floodplains or via the *hyporheic zone* of seepage down the valley floor. The same channel may in fact recharge groundwater in floods and receive contributions back during droughts; this interplay is of extreme importance for water chemistry and freshwater ecology.

Clearly the very wide range of porosity and permeability in rocks makes their contribution to baseflow in rivers equally variable. In order to predict baseflows in ungauged catchments a number of attempts have been made to express this sustainability factor; how long can low flows of a certain probability of occurrence be kept going from the aquifers of a catchment? The UK Institute of Hydrology (1980) have developed a baseflow index (BFI) for catchments calculated as follows:

> The annual hydrograph for a catchment is separated into its flood peaks and baseflow (as illustrated on Fig. 3.3). This procedure can be carried out semi-automatically. The BFI is the proportion of the total outflow accountable to baseflow over the period of record. Table 3.2 shows that in the UK it is highest in chalk and other porous limestones and lowest in clay lithologies. The BFI can be derived from as little as one year of flow data and is a powerful prediction of other aspects of the low-flow performance of catchments (see 4.5.2).

3.3.4 Applied groundwater hydrology and environmental problems

Because this book takes an environmental theme in hydrology we now introduce a brief sample of environmental problems involving groundwater-flow processes. First, to illustrate the 'simple' *fluctuation of water-tables*, and its impacts, we consider the exploitation of groundwater for water supplies.

In the United Kingdom the exploitation of groundwater began on a large scale with the harnessing of power to pump from depth and as a result of the damaging epidemics of the nineteenth century (largely spread in the surface water system). Beneath large cities such as London, Birmingham, and Liverpool water-tables fell until some sort of equilibrium was restored between recharge and abstraction. This decline of water-tables has continued in some locations near to the coast, and sea-levels now stand higher than adjacent groundwater tables. The result is *saline intru-*

sion and a drastic decline in water quality for supply; eventually such boreholes need to be abandoned. In our largest cities supplied by local groundwater, borehole abandonment, largely because of declining inner-city populations, has also had a damaging effect. Engineering structures including foundations and railway tunnels, built during the heyday of pumping are now threatened by *flooding from rising groundwater*.

Fluctuating groundwater levels call to mind those of a reservoir and the positive utilization of aquifers as storage-below-ground has been a feature of water management in many nations in the last thirty years. For example, in south-east England the important chalk aquifer is used to store surface water by *artificial recharge*. Purification is applied

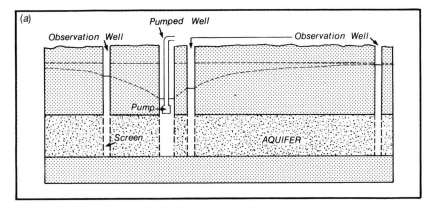

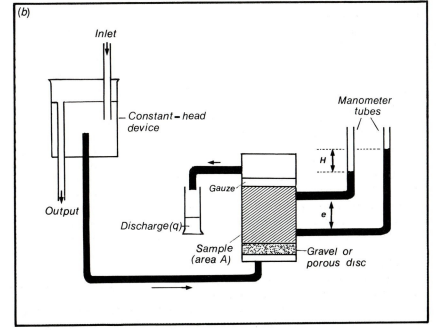

Fig. 3.6. Testing for values of saturated hydraulic conductivity. (a) by pump tests on groundwater systems in the field; (b) by laboratory tests of soil or rock samples.

In (a) the dashed lines represent the original and pumped water tables. When compared with the pumping rate or recovery on cessation permeability can be calculated. In the laboratory (b) water is 'forced' through the rock sample, producing an analogous situation.

Table 3.2. Typical baseflow indices for various rock types in UK

Dominant permeability characteristics	Dominant storage characteristics	Example of rock type	Typical BFI range
Fissure	High storage	Chalk	0.90 – 0.98
		Oolitic limestones	0.85 – 0.95
	Low storage	Carboniferous limestone	0.20 – 0.75
		Millstone Grit	0.35 – 0.45
Intergranular	High storage	Permo-Triassic sandstones	0.70 – 0.80
	Low storage	Coal measures	0.40 – 0.55
		Hastings Beds	0.35 – 0.50
Impermeable	Low storage at shallow depth	Lias	0.40 – 0.70
		Old Red Sandstone	0.46 – 0.54
		Silurian/Ordovician	0.30 – 0.50
		Metamorphic–Igneous	0.30 – 0.50
	No storage	Oxford clay ⎫	
		Weald clay ⎬	0.14 – 0.45
		London clay ⎭	

Source: Institute of Hydrology (1980).

before recharge to avoid the risk of damage to the water quality of existing reserves; recharge is also practised in Germany and in the coastal sand-dunes of the Netherlands (Kirby, 1984). A reversal of recharge can be practised where surface streams need replenishment for ecological and fisheries purposes (Wright and Berrie 1987).

Elsewhere in the world there are no surface-water resources in surplus with which to carry out artificial recharge; natural recharge from precipitation is little or none (Lloyd, 1986) and continued exploitation of groundwater is equivalent to mining. The accurate calculation of recharge to aquifers in such climates is critical; some nations are assuming that because the water balance has no surplus for recharge toxic wastes can be dumped in landfills—a further threat to future sustainable use of this finite groundwater resource (Gee and Hillel, 1988). The Ogallalla aquifer of the High Plains of the USA is 'mined' by six states, by an equivalent depth of 2 m per year, whilst rainfall is recharging it by 12 mm per year (Reisner, 1990).

An interesting variant of groundwater mining, with a political, competitive element occurs in the Middle East. Israeli settlers in the West Bank are sinking deep wells which, when pumped, create a depression in the water-table sufficient to seriously deplete the Arabs' traditional tunnel supplies from the same groundwater resource (for details see Rowley, 1990).

Groundwater pollution is often undiscovered in public water supplies until well after the surface event has occurred; pollution may even have become habitual as in the case of the disposal of waste (including hospital wastes and cleaning fluid) into soakaways and the heavy use of fertilizers and pesticides by agriculture. Rivett *et al.* (1990) reveal the extensive pollution from chlorinated solvents (mainly arising in metal cleaning) to the Triassic sandstone aquifer beneath Birmingham; fortunately it is not used for human supplies but their survey reveals 'proofs' of pollution near the base of the aquifer and above zones of less permeable strata (minor aquicludes). Clearly, whilst pollutants are slow to enter aquifers they are also slow to disperse and may even accumulate; the advantages of groundwater as a source of supply must not be compromised long-term by lack of protection and management. Aquifer protection should be a policy aim of pollution control everywhere; the example of Nitrate Sensitive Areas is developed in Section 8.6.

3.4 Soil moisture

In contrast to the global contribution of groundwater to freshwater supplies, soil moisture in the root zone of plants constitutes only 0.064 per cent of fresh water; however, because it feeds the roots of plants it has received a great deal of attention from scientific hydrology.

Most of the principles we shall apply to moisture in the soil are identical to those already established for groundwater flow. However, soil water is notable mainly because of a much more direct and widespread human intervention in its control, e.g. through irrigation and drainage. Soils differ from rocks in the following ways which influence the boundary conditions for subsurface flows and again help to make soil water a distinctive field of study:

(1) Soils are very variable both laterally and vertically; they also change through time—slowly by weathering and erosion but seasonally through plant growth or cultivation,

(2) Unsaturated zone dominates soils (cf. groundwater) since this is the management zone; soil water-tables are often temporary and are of interest mainly to drainage engineers.

(3) Soils are less homogeneous in texture than most rock types and pores adopt a large variety of sizes (see Section 3.4.1).

(4) Soils have a high organic content which helps make them much more chemically reactive than most rocks; the chemistry of soil water is therefore much more unstable than that of groundwater.

3.4.1 Pores, permeability, and pressure

To even a non-technical eye most soils are porous, their ability to store and to throughput water is obvious. However, few will have realized the nature of soil porosity as revealed by the lens or microscope (Fig. 3.7). This figure reveals two aspects: the existence of larger *macropores* which are very important in water transmission, and a general variability of size and location of all pores. The soil depicted is unlikely to perform as the medium envisaged in Darcy's Law, i.e. as an *isotropic* medium, able to transmit water in any direction according to the

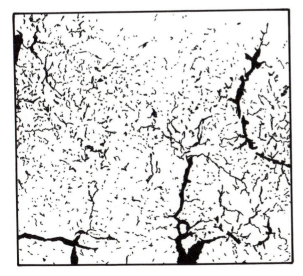

Fig. 3.7. Pore spaces in a soil section seen under a microscope.

governing head and hydraulic conductivity. A further complication is that permeability will not only vary with the variability of pores, producing *anisotropic*, concentrated flow behaviour, but also with soil moisture content under unsaturated conditions. Since soil saturation and therefore soil water-tables are relatively infrequent in the upper layers we are often dealing with the prediction of unsaturated flows.

Soil permeability is complicated by the development of forces additional to that of gravity; they arise as a result of capillary movements between grains and the additional effects of surface tension on water films around grains. In addition to gravity potential there is therefore a *matric potential*, derived from the soil matrix. It is negative (it acts to retain moisture) at moisture contents below *field capacity*, i.e. at the end of gravitational drainage. As Figs. 2.11 and 2.15 show, for any given soil, moisture is still available to plants against the surface tension (because roots develop their own forces) until *wilting point* is reached. As the figure shows, soils vary considerably in regards to this phenomenon—a factor of great significance to properly managed irrigation and drainage.

We are able to measure soil-moisture tension with *tensiometers* (see Section 2.7), the simplest comprising a water-filled porous pot through which the tensions in the soil are transmitted, feeding this information to a water or mercury column. The rise and fall of the column reflects the positive (> field capacity) or negative (< field capacity) matric potentials, yielding a measurement which can help us predict movements of moisture. Thus to predict moisture movements on a slope we need to measure the gravitational head, above for example the stream, and the matric potential to create the contours governing flow. It is therefore necessary to have a generous number of tensiometers, preferably recording automatically, at a number of locations determined by slope topography and at a number of depths to reflect soil horizons. Anderson and Burt (1978) used this technique to map saturated areas as a dynamic feature of hillslope hollows.

We are still assuming that the lateral permeability (hydraulic conductivity) figures featured in Fig. 3.5 can be applied to field soils. In soils we are, however, additionally interested in the vertical dimension. The vertical conductivity of soils controls the *infiltration rate* (whose maximum is termed the infiltration capacity). Field tests to determine this rate and capacity are particularly important in regions receiving intense rainfalls since the excess of

Fig. 3.8. Construction of troughs to collect and measure throughflow from a soil section (side view).

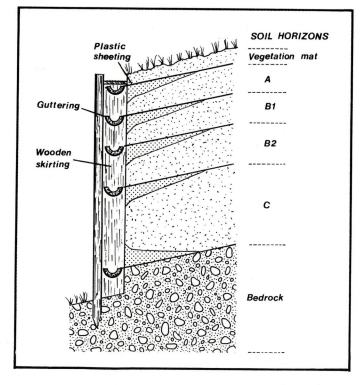

rainfall intensity over infiltration capacity represents the potential to generate runoff by rapid surface routes. The *cylinder infiltrometer* (Plate 3.1, foreground) is one type of equipment designed to measure infiltration capacity—a known head of water, supplied from a reservoir, is applied to the soil surface within a cylinder to restrict moisture movement to a vertical plane. An alternative, more realistic strategy, which has additional utility in soil erosion studies, is the *rainfall simulator* (Plate 3.1, background) which delivers realistic droplets at a known rate, from a pump and nozzles, to the ground surface within a demarcated plot; runoff is collected from the plot and its rate is subtracted from (artificial) rainfall intensity to give infiltration rate (in mm h^{-1}).

3.4.2 The auger-hole method

Moving on to horizontal permeability of soils it is essential to highlight a particular field technique for the determination of saturated hydraulic conductivity of soils because of its widespread deployment in both agriculture and in the conservation of wetlands. The auger-hole method is the equivalent in soil hydrology to the pump test (Fig. 3.6(a)) in groundwater studies.

There are four simple steps in the field:

(1) A hole is bored in the soil using an auger (commonly *c*.50 mm diameter), to the depth of the horizon whose conductivity needs to be measured.

(2) The auger hole is normally lined, to prevent collapse, with plastic tubing perforated to admit water only from the chosen horizon.

Plate 3.1. Techniques for measuring infiltration—by rainfall simulation (left) and cylinder infiltrometer (right).

(3) Once a water level has developed in the tube or piezometer (since it is the piezometric head which is measured, though in many soils there are no confined aquifers)—it is pumped out.

(4) The test consists of timing the recovery of the water level to its previous height in the tube.

Clearly, what is in effect being measured, from data on piezometer dimensions, is the discharge from the chosen horizon laterally into the piezometer. Since the head powering the discharge and the area across which it is occurring are known, the conductivity (k) is calculated as the only unknown. The full technique is described in detail, with illustrative calculations by Smedema and Rycroft (1983).

3.4.3 The special case of peats

Section 7.8.1 reveals the peculiarities of soil hydrology in wetlands; hydraulic conductivities are known to be low but for practical conservation projects it is often necessary to know 'how low?' The auger-hole method is difficult to apply in wet peats—smearing of the sides of the hole occurs and the withdrawal of semi-liquified peat from the hole poses further problems. However, using an extremely careful field technique Ingram et al. (1974) successfully carried out some eighty tests on peats at Dun Moss, a raised bog in Scotland. The authors were surprised to find that rates of recovery of the water level in their piezometers varied continuously through time: no one rate could be selected for the calculation of conductivity. k varied from 2.10^{-4} cm sec^{-1} to 2.10^{-4} cm sec^{-1} for one auger-hole piezometer. It is possible that gas trapped in the extremely fine pores of the peat mass interferes with the movement of water. Removal of peats to the laboratory for permeameter tests is virtually impossible without severe disturbance; we therefore remain in considerable ignorance of a basic parameter in the hydrology of one of our most sensitive soils.

3.4.4 Soil-moisture patterns and flow

To the geographical hydrologist working in the field there are severe problems in adopting the detailed principles of soil physics developed in broad scope above; however, there are severe penalties for ignoring them. Thankfully, conventional objects of the field-worker's attention such as soil mapping and vegetation patterns offer considerable amounts of information which extend the principles of soil physics to the actualities of soil-moisture status and movement, especially on slopes.

Major elements of 'the real world' to the field observer are:

(1) The clear development of *soil horizons*, each with its own properties of porosity (particularly macropores), with distinct boundaries representing rapid changes in permeability between these horizons.

(2) The influence of *geomorphology* on soil-moisture movements; even microrelief can influence flow routes and there can be positive feedbacks since seepage zones, once established, can promote more rapid bedrock weathering—exaggerating the microrelief.

(3) The role of *macropores* developed as the result of drought (cracks), animal activity (burrows), plant activity (root channels), and the particular role of soil pipes (see Section 3.7.4).

(4) The role of *vegetation*, natural, semi-natural, or crop in controlling moisture contents and flow routes (especially where a crop or plantation has cultivation, drainage, or irrigation).

(5) The role of *zones of temporary or permanent saturation* under conditions of low rainfall intensity (i.e. much of the humid-temperate zone), in controlling runoff processes in the soil and hence soil-water quality.

The latter point will be developed in subsequent sections of this chapter but, in relation to points (1) to (4), the influence of soil horizons, microrelief, and macropores has largely been elucidated by quite local studies involving a variety of soil pits, water-collection systems, moisture-measuring systems (both level and tension), and mapping of saturated areas, all within small, often forested, catchments. Fig. 3.8 illustrates some of the techniques. Atkinson (1978) provides a review. Forested catchment studies have often driven along the pace of hydrological innovation and the discovery of patterns of soil-moisture movement on slopes was one outcome of the intensive research effort in North American forests in the 1960s (see Hewlett, 1982).

It is fair to say that other soil and slope conditions have been relatively neglected, particularly flat, agricultural land; however, it could be argued that artificial controls to runoff are always applied to such areas and scientific hydrology must find, as far as possible, undisturbed conditions in which to work: hence wilderness and natural forests.

In the United Kingdom a good deal of research

Fig. 3.9. The basic hydrological questions posed by design of a simple agricultural under-drainage scheme.

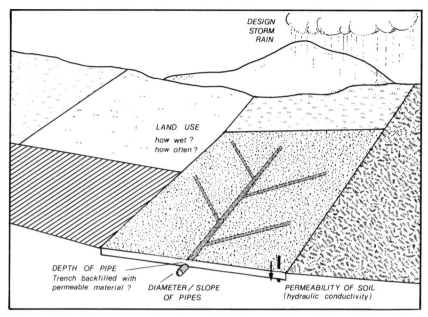

progress has been made under *moorland* vegetation and soil covers. Here soil morphology and chemistry reflects the drainage (particularly saturation) status of soils and the semi-natural vegetation of the shallow interfluves, steep slopes, and depositional valley floors sets out the patterns of water movement. Newson (1979) describes the use of both soil and vegetation maps in the spatial demarcation of catchment 'runoff domains' in distributed mathematical models. Box 3.1 lists these domains for an upland catchment in mid-Wales, UK.

Working in lowland Britain, Gurnell, Hughes, and Edwards (1990) separate five categories of vegetation which reflect both soil-moisture contents and the regularity of surface saturation. In turn the combination of vegetation types at catchment level has a direct bearing on the proportion of rainfall which becomes storm runoff. Clearly topography acts as a gravitational control on moisture in the first place but there are management implications for the way the vegetation cover is treated, especially in respect of interception losses.

3.4.5 Applied soil hydrololgy

In immediate contrast to the preceding section one must say that studies of soil-moisture movement in field soils as yet represent only a small part of applications in this topic. Soil-moisture contents and tensions are, however, critical in the management of soils in two major categories of application:

(1) Soil *drainage* by gravitational forces, down to field capacity.

(2) Soil-moisture *availability* to plants between field capacity and wilting point(s)—and how to restore dry soils as a plant substrate.

Latterly, however, we have had a fast-developing third category of application:

(3) Soil-moisture *movement* between land requiring contrasting water regimes, particularly between agricultural or urban land and wetlands.

Turning first to drainage, its principles as a management activity long pre-date scientific hydrology. The Romans drained land as they expanded into oceanic climates where soil saturation and surface flooding halted the spread of their civilization. Later, patterns of ploughing clay soils ensured a network of routeways across the surface to collecting channels. However, after subsoil drainage for farmland was introduced, first as buried lines of stones, then as pipes and recently as pipes-plus-stones, our need to predict their effects on soil moisture made the exercise much more technically demanding.

The basic problem is determining flows to subsurface drains as set out in Fig. 3.9. Because drainage seeks to improve crop productivity the aims of the process must be clear; one begins, therefore, with a choice of grassland, cereals, horticulture, etc. and proceeds to the choice of a pipe size (diameter) for the scheme—flow is an intermediate calculation *en route* (see Ministry of Agriculture, Fisheries and

Box 3.1. Runoff domains in upland humid-temperate landscapes, based upon field-process measurements

Landform and origin	Mineral cover controlling processes	Vegetative/structural cover indicating processes	Current morphological trend
Tertiary plateaus *runoff feeds into* →	Blanket peats *runoff feeds into* →	Wet moorland *runoff feeds into* →	Rapid peat erosion— to bedrock →
Glacially/periglacially modified valley-sides *runoff feeds into* →	Pedogenic peat/peaty soils *runoff feeds into* →	Drought-tolerant grasses *runoff feeds into* →	Piping, gullying, slope failure extends channels *runoff feeds into* →
Glaciated depositional valley-floors *runoff feeds into* →	Gleys *runoff feeds into* →	Soligenous mires *runoff feeds into* →	Channel migration, bluffs *runoff feeds into* →
Incised channels — to bedrock →	Bedrock/boulders/gravel	Falls and pools, bars	Transport/storage/bank erosion

Food, 1983). Flow from a field is related to:

(a) the rainfall of a chosen probability (in mm d^{-1});
(b) hydraulic conductivity of the soil;
(c) catchment area to each drain (depends on spacing of drains—spacings around 10 m are common);
(d) pipe gradient.

Normally drainage agencies simplify the procedure into a series of nomographs (more recently computer 'expert systems') which ease the chore of calculation. Recent field trials have proved that pipes alone seldom promote efficient drainage—some degree of secondary treatment to crack the surrounding soil, or network it with unlined 'mole' drains is essential to conduct water to the drainage pipes.

Only brief consideration of applied soil hydrology in the field of irrigation can be made in a chapter about runoff processes. Rather than treat the technical specification for the addition of soil moisture through irrigation (see Section 4.5) we may briefly deal with the method of *soil-moisture accounting* by water balancing which is used to define moisture stress over wide areas (direct soil measurements are seldom part of hydrometric networks outside research catchments).

A moisture account may be set up for regional soils on a seasonal or daily basis; the seasonal occurrence of periods of moisture deficit (evaporation exceeds precipitation) is a diagnostic feature of most world climates (see Fig. 3.10(a) and (b)). Regular measurements of precipitation and estimates of potential evapotranspiration are the basis for these plots and they significantly constrain agricultural opportunities (both of crop choice and yield)—see Jackson's review of tropical conditions (Jackson, 1989).

Irrigation within the season of deficit requires, however, a much more sensitive day-to-day knowledge of the moisture account and Fig. 3.10(c) illustrates the method of calculating deposits and withdrawals which is analogous to those to and from a bank account. Thus, UK irrigators can subscribe to bulletins of Soil Moisture Deficit from the Meteorological Office (see Section 4.5.2). The actual modelling of deficit requires a more complicated accounting system; for example, meteorological estimates are of potential evapotranspiration—not actual—and root constants need to be incorporated. Clearly, too, the relationship between moisture content and moisture tension needs to be known for

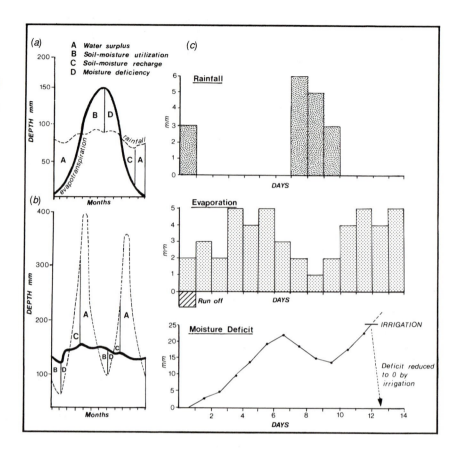

Fig. 3.10. Soil Moisture Deficits and irrigation scheduling. (a) The annual pattern for a humid-temperate location. (b) The annual pattern for Colombo, Sri Lanka (modified from Jackson, 1989). (c) A simple temporal water account and the effect of carefully estimated irrigation water.

each major soil type. Finally, when irrigation water is added from above there will be interception losses on the crop canopy. One of the biggest challenges to agricultural productivity in the next century will be efficient utilization of irrigation water, including the use of drainage systems to remove any build-up of salts in irrigated soil columns.

Irrigation systems are not widely employed to conserve wetlands (Chapter 7) but we have already referred to the need to maintain soil moisture (often saturation) under conservation areas whilst permitting productive agriculture nearby. Simplistically we may apply the same principles of soil physics used to design drainage networks to calculate soil moisture flows between two such adjacent sites: these calculations may then be used to fit protective membranes, dig boundary canals, or establish buffer zones between the wetland and the farmland or forest.

3.5 The ground surface

To the field observer ground-surface properties are an obvious *variable* in hydrological-process studies rather than the *constant* to which they may be assigned by simple models. Even a desert sand surface may be granular or crusted and is corrugated in various ways by wind. More spectacularly, the humid agricultural land surface may be crop-covered, crop-waste covered, cultivated, tracked by vehicles, frozen, flooded, or cracked.

The geometry of the ground surface exercises a control on the routeing of rainfall, rough surfaces promoting surface *detention* and/or *infiltration* and additionally resisting the downslope flow of surplus rainfall. Smooth surfaces tend to promote surface flows but these in turn, if erosive, roughen the surface. The literature reveals much more on the erosional effects of different configurations of the ground surface; clearly management practice in cropping systems can be used to ameliorate the loss of soil. In pure hydrological terms the conclusion of most research appears to be that surface conditions are closely related to the ability of rainfall to infiltrate and it is this latter property which controls runoff routing. Thus a smooth ground surface tends to be a hard surface and a cultivated rough surface often bespeaks an artificial drainage system whose influence is the greater.

For the ground surface to play a major role in runoff processes the soil 'tank' must obviously be either:

(a) unable to conduct rainfall or snowmelt to depth—*insufficient infiltration capacity*; or
(b) full, i.e. *saturated*, water being impermeable!

The earliest model of surface-flow generation relied heavily on the conditions described in (a) above; R. E. Horton (1945) worked in those areas of the United States where rainfall intensity regularly exceeds infiltration capacity and Fig. 3.11 indicates how the rainfall excess first becomes detained by surface hollows and then forms a flow of increasing depth downslope. Horton's contribution was immense because he linked the hydrological system to river-basin geomorphology (the surface flows perform erosional work); the United States also suffers from considerable problems of soil erosion for which the model provided the prospect of rational ameli-oration.

Horton's model is recognized by the use of his name to describe surface flows generated in this way as '*Hortonian overland flow*'; however, the model does not hold good for regions of the world with more permeable soils and less intense rainfalls. Here, as will become clear in Section 3.7, it is soil saturation which promotes surface flows and the resulting process is labelled '*saturation-excess*' *overland flow*.

Hydrological models have little practical need to separate the two forms of overland flow and they tend to be lumped as 'quickflow'; however, as interest grows in the chemical signature of runoff from different sources it may well be necessary to separate what is the rainfall chemistry of Hortonian overland flow and the soil influence on the chemistry of runoff from saturated areas, particularly where saturation forces throughflow to the surface (see Section 3.5.3).

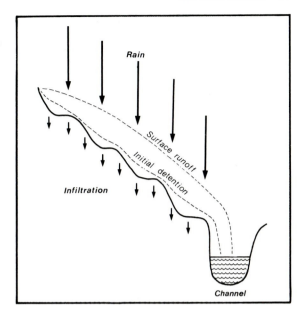

Fig. 3.11. R. E. Horton's exaggerated depiction of runoff from slopes (emphasizing overland flow).

3.5.1 Roughness: cultivation and vegetation

In humid regions there is another factor affecting the hydrological importance of ground surfaces, i.e. the perennial or seasonal cover of a vegetation canopy. Whilst this cover may mask the soil surface more or less thoroughly it sets up a control on the rate at which *net* precipitation reaches the ground and also controls the pattern and energy of impact on the ground (see Chapter 2). When overland flows are generated, whatever their origin, their route downslope will often be across a surface on which vegetation (stems/trunk elements) is the main factor contributing microrelief. Vegetation debris, contributing the litter layer and upper-soil horizons, may be responsible for much of the near-surface transmission of runoff which is effectively, therefore, operating as a porous medium.

Truly overland flows are analogous to the chan-nelized flows described in Section 2.5.1 but occur as very shallow layers over, proportionally, very rough surfaces. The list of Manning's n data provided in Table 2.5, when extended to the floodplain, indi-cates the frictional effect of such vegetated or culti-vated surfaces. Very few direct measurements of overland flows have been made because, as we con-clude in the following section, much of our atten-tion has been given to 'smooth' surfaces for overland flow.

The main components of *frictional resistance* to surface flows include those produced by soil particles, microrelief, changes in flow regime (small standing waves), and, because the depths are so small, raindrop-impact turbulence. As in the case of channel flow, resistance decreases with flow and indeed *sheet flow* rapidly agglomerates into small channels called rills whose hydraulics are those of normal turbulent flows (sheet surface flows may remain non-turbulent, i.e. *laminar*). Storage plays a considerable role in the hydrology of cultivated surfaces, water collecting in furrows; clearly in erosion protection of base soils cultivation patterns can maximize storage and minimize the compaction, e.g. along wheeltracks which can pre-form rills and eventually gullies—through soil erosion.

3.5.2 'Smoothness': urban and paved surfaces

Hydraulically smooth surfaces abound in urban areas—housing areas and transport networks are finished to a smooth and often impervious surface. However, urban areas are also equipped with surface drainage systems to cater for excess rainfall; once again it is difficult to disentangle the hydrological effects of the surface from that of the accompanying water-management systems. Studies have only recently begun of the surface effects alone (Hollis and Ovenden, 1988); these reveal that *infiltration* into 'hard' urban surfaces, such as roads, is much greater than previously assumed. The earliest predictive models to guide the design of urban drainage systems assumed that 100 per cent of rainfall would become runoff across urban surfaces. Modern design is increasingly treating the same surfaces as capable of permitting infiltration—innovative materials have been used in places with this in mind. It should also be realized that urban land cover includes a considerable component more typical of the rural zone: parks, gardens and trees.

Hydraulically smooth urban surfaces have also been assumed to accelerate the flow of the large amounts of runoff they produce, compared with velocities observed on natural surfaces. This is true up to a point but in the previous section we observed that frictional resistance to shallow overland flows included very small particles, flow transitions, and raindrop turbulence. Few surfaces in urban areas are truly smooth but they are planar; thus transitions to rill-flow are more unusual. Standing waves commonly develop on urban surfaces. The importance of raindrop turbulence is revealed by the fact that discharge from urban surfaces can temporarily increase after rain ceases.

Storage is very important on paved surfaces; urban areas are notorious for puddles, flooded roads, car parks, etc. This storage potential means that the rapidity of flow from some paved surfaces is negated in the overall time performance of a paved catchment. We shall develop this theme further in Chapter 5.

3.5.3 Surface-flow patterns: sheets, rills, and 'springs'—return surface flows

Flows across both bare and vegetated surfaces develop certain patterns according to their mode of generation and in relation to microtopography (with which there are, of course, positive feedbacks if surface flows achieve erosion of that surface). Hortonian overland flow, in theory, develops as a thin sheet because on uniform slopes an excess rainfall intensity will exceed infiltration capacity relatively evenly and almost simultaneously. Sheet flows can occasionally be observed, especially on desert surfaces and in eroding claylands; in humid-temperate conditions it can be seen during exceptional rainstorms especially on compacted farmland or in urban areas.

Flow in sheets is much rarer where it is generated by saturation of soils (Plate 3.2). Patterns of saturation tend to be guided by the slow drainage of gravitational moisture between storms and are hence related to topography, saturated areas corresponding with the base of slopes, especially in hollows where drainage converges (Plate 3.3). Hortonian flows also rapidly break down from sheets into more concentrated flow lines, effective in creating gulleys in erodible surfaces. A further type of concentrated flow is *return surface flow*, originally throughflow (having infiltrated the soil upslope) which gravity forces to the surface on meeting the impermeable wedge of saturated conditions, e.g. at the foot of slopes. Return surface flows may occur as a zone of pronounced seepage or commonly as small springs resulting from the focusing of throughflow in macropores.

Whilst concentrated surface flow routes may not open up discernible channels, especially in vegetated surfaces, they clearly act as extensions to the channel network during floods and the reader is encouraged to conceive of the runoff system as highly dynamic in this way (see Section 3.7 below).

Plate 3.2. Sheet flow converging and dispersing at the foot of a saturated pasture—a relatively rare site in the United Kingdom on thickly vegetated soils.

3.6 The stream channel

Topologically the network of open-stream channels fills the humid land surface with certain regularities of pattern. Stream networks are dendritic patterns: they have a measurable *drainage density* (length of channel per unit area of catchment) and the reciprocal of this index (the *constant of channel maintenance*) indicates the runoff area supplying each unit of channel. Clearly the two concepts do not match: the supply of runoff to unit lengths of channel holds true only in fingertip tributaries (*first-order*

streams—see Strahler, 1957) whereas the dendritic network implies the gathering function of a system of channels of different capacities. Lower down in the network the conveyance function dominates the collection function in terms of catchment runoff. In Section 3.7 we are made aware that zones of runoff production can also be identified within catchments.

Experience tells us that, topographically, stream channels are very variable in both planform and section; more importantly for hydrologists, channels may be classified on the basis of their relationship with the valley they occupy.

Plate 3.3. Techniques to indicate surface runoff (left) in relation to soil saturation (right). The scale is marked in inches.

3.6.1 Channel classification

In connection with overland flow we observed that it is difficult to separate the point where porous flow in the root zone becomes a thin film of sheet flow. The image of a dynamic continuum in runoff processes must be maintained into channel classification where the smallest end-members may be ephemeral and not open to the surface (Table 3.3).

Manning's *n* offers a simple principle for channel classification (Table 2.5 encourages this view) but since its application is more appropriate to the condition of short reaches the classification of channel networks is more logically based on simple descriptions as in Table 3.3; this classification, appropriate to small upland basins, was developed as part of a modelling strategy. Clearly in much larger basins a less detailed system is required and it is unlikely that maps of the type shown in Fig. 3.12 can be

sustained (unless remote sensing techniques can be deployed). Geomorphologists have developed relatively simple techniques of channel classification for large basins (see Table 3.4), though they emphasize the sediment transport rather than the runoff function of the network. We now need to reverse this priority with some detail as to how flow builds up in a channel network of the type shown in Table 3.3.

3.6.2 Channel processes: transport of water and sediment

Stream channels are formed by the processes of water flow and sediment transport; whilst chemical solution is of major importance in weathering, its activity in channel formation is fairly minimal outside limestone terrain. For many years the approaches of both engineers and geomorphologists to stream-channel geometry has been one of

Table 3.3. Classification of Plynlimon channels

Class	Regime	Type	Description
Closed channels	Artificial	Tile drains	An unrepresentative absence in the Wye catchment.
	Intermittent	Soil pipes	Flow only during and after moderate or heavy storms. Common midslope in the Hiraethog/Hafren profile.
	Perennial	Flushes	The perennial pipes of Gilman-bedrock depressions with peat infilling; *Juncus effusus* most useful indicator.
Open channels	Intermittent	Ephemeral channels	Includes the extending network of the eroded peaty plateaus and the grassy 'gullies' of Bell and Calder. Both types may initially have been formed during previous cover/climate conditions. Formed in unconsolidated superficial deposits.
	Perennial	Peat–lined channels	The upper channels of both Wye and Severn main streams flow between peaty banks. Channel bed is often in peat. Active erosion represents failure of peat to roof over stream completely.
	Perennial	Bedrock channels with falls and pools	Particularly common in upper reaches where channels cross the strike of the local rock. Flow path swings criss-cross to channel direction.
		Boulder channels with alluvial banks	Glacial dumps of boulders occur prominently at base of steep fall-and-pool reaches.
	Perennial	Alluvial channels	Banks are peat and silt on cobbles. Bed of cobbles and gravel. Meanders, riffles, and pools.
	Artificial	Drainage ditches	Dominant in the middle part of the Severn catchment in deep peat. Some old open drains in Wye, mainly vegetated over.

Table 3.4. Stream-channel reach classification for large river basins

Zone 1	GORGES and undercut slopes contribute directly to channel but in fact relatively little eroded material added except in slope failures.
Zone 2	ARMOURED with gravel beds firm except in events of infrequent occurrence when armour is disrupted. Bank erosion more frequent.
Zone 3	TRANSITION from armoured to sand, with a fine load which can be moved more frequently and which forms erosion-susceptible banks (once any vegetation cover breached).
Zone 4	SAND or MOBILE reaches with mobile fine gravel or sand bed; transport at most high flows. Sand-bed streams are distinctive, with very long mobile reaches; rapid lateral erosion.
Zone 5	BACKWATER: formed by reduction in gradient and increasing cohesion of fine sediments; adjustments very slow and overbank flows deposit on (and form) floodplain.

Source: Modified from Warner (1987).

equilibrium—the channel network, its components, and individual reaches having a 'nice adjustment' to the system's function as a water and sediment conveyor. Such models have assumed that the well-proven correlation between full-channel (bankfull) discharges and optimum work-rate also implies this particular flow as dominant or channel-forming. However, more recently geomorphologists have also researched threshold models of channel formation which give a formative role to extreme floods; these models have been found particularly appropriate to upland rivers or other channel reaches controlled by coarse bedload. For a review of these concepts see the chapter by Lewin, 1989.

Retaining an equilibrium approach for this Section, it is helpful to consider the alluvial channel (i.e. one formed in river-transportable materials) as having the following adjustments available in the 'meshing' with formative flows. For the moment let us assume that these flows are at bankfull and that they occur approximately annually. The degrees of freedom for adjustment are:

width
depth
slope
velocity
sediment calibre
channel planform.

In practice each element of the list has a characteristic spatial and temporal scale of adjustment. For short- to medium-term considerations attention has focused on width, depth, and velocity adjustments; engineers have formulated *regime equations* for these parameters, with the formative discharge as predictor whilst geomorphologists have calibrated the *hydraulic geometry* of alluvial channels at a range of discharges (see, for example, Richards, 1982).

The most instructive of the depictions of hydraulic geometry relations was provided by Leopold *et al.* (1964). Their graphs, shown here as Fig. 3.12, illustrate the way in which width, depth and velocity react to increasing and decreasing flows both *at a station* (i.e. at one point through time) and *downstream* (at different points in the network under similar conditions).

It is particularly useful for those who have learned at school that upland rivers are 'fast' and lowland rivers are 'slow' to look at the variability of velocity at a station and downstream. In a clear demonstration of an adjustment between slope and hydraulic roughness, downstream variability in velocity is limited (a point proved in the field using current meters by Ledger, 1981).

Other investigations of flow through channel networks have used tracer (dyes or chemicals) to determine network velocities under varying conditions (see Beven *et al.* 1979). Network velocities vary strongly with discharge (from <0.1 m s^{-1} to 1.0 m s^{-1}) but also with the geometry of individual reaches; those with fall and pool sequences showed much lower velocities than straight or meandering types.

In terms of sediment transport in the channel network, geomorphologists have become increasingly aware of the importance of sediment *supply* to explain its variability. Broadly there are zones corresponding to a downstream-fining sequence (cobbles—sands—silts/clays) but supply is not merely from the catchment—it is also a function of channel margins. Thus, whilst silt/clay material in suspended load may pass through the channel network quite rapidly, the movement of coarser material tends to be restricted to floods, to zones where it is available, and occurs as a series of hops between storage zones.

3.6.3 Channels and river valleys

Channels are clearly part of the hydrologist's purview: they contain and control the hydraulics of flow from a catchment. Their slope and capacity adjust over time to the flow but this adjustment may be neither rapid nor smooth. The channel's

Fig. 3.12. Hydraulic geometry: the contrasting changes to channel width, depth, velocity, and suspended sediment load with flow at a site (A–C and B–D) and in the downstream direction (A–B and C–D) (after Leopold *et al.*, 1964).

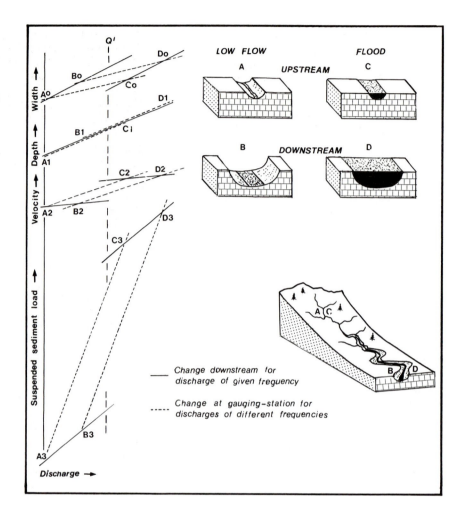

relationship with the valley floor subtly influences the nature of this adjustment. For example, an accepted definition of a flood is flow out of banks (afflux). However, in two adjacent channel reaches, one *depositing* sediment in the bed, and the other *incising* the river bed, an identical high flow will or will not (respectively) constitute a flood. The valley-floor sediment supply may well control the location of such aggrading and incising reaches. In the short term, during one flood-flow any given channel section may first incise (scour) as velocities increase and then aggrade (fill) as they fall but, considered as a whole, the channel 'conveyor belt' for sediments is jerky.

3.7 Assembling the runoff cascade: a model

Let us first examine the practical needs in environmental management for a working understanding of the ground phase of the hydrological cycle. Traditional applications of hydrology as we have seen and will develop in Chapter 4, are, in the main, content with bulk, lumped, black-box approaches to the catchment, often statistical in their nature. We need to have a long set of data on flood-flows or rates of catchment losses upon which to base these statistical predictions but they have a long record of success. Understanding the detail of processes strengthens confidence in the predictions but this detail remains an implicit check rather than an explicit function of the predictions.

However, in three situations, familiar in environmental hydrology, use of detail is necessarily explicit:

(a) Where data are absent or scarce, requiring a 'cold start' prediction from a deterministic predictive model of runoff generation.

(b) Where a pollution problem pervades the process cascade, as does acidification, beginning with acid precipitation.

(c) Where land-use changes have a polluting effect (e.g. fertilizer runoff) or where the change modifies the evaporation and soil-moisture properties of the system—a pervasive effect through the cascade.

3.7.1 Slope and channel linkages

Slopes supply channels with runoff, solutes, and sediments; however the linkage is by no means direct or uniform. In the uplands the relationship between slopes and channels is immediate, often spectacular (e.g. river cliffs or scree slopes). Both surface and subsurface flows of water have almost direct gravitational access to the channel margins (Plate 3.4(a)). However, a variety of other slope–channel linkages exists throughout the channel network (Table 3.4 and Plate 3.4(b)).

We have referred before to hyporheic flows below and around surface channels and to the variable recharge–discharge relationship between surface and groundwater flows. Environmental hydrology has only recently begun to elucidate the phenomena of the riparian zone; Table 3.5 is a highly simplified summary of a situation which has a key influence on stream chemistry, pollution control, and stream biota.

An important single lesson from this Section is, however, that throughout the majority of larger river basins the contribution of runoff by surface, soil, or groundwater routes to open channels occurs through a further 'tank' in the cascade—the *valley floor*, *riparian zone*, and channel margins or *river corridor*. The dilemma of the field experiment in hydrology is once again exposed—whilst the most controllable experimental work is carried out in small, headwater basins or on slopes, the assembly of a model for larger basins is not merely the sum of the parts. A great deal more work is required on the highly dynamic zone of the river valley floor; we return to scale problems in Section 3.7.5.

Plate 3.4. (a) Slope-channel coupling in typical headwater source area. (b) Slope-floodplain channel coupling of transfer zone.

3.7.2 Partial contributing areas

Very little of the foregoing material on runoff processes has encouraged the impression that they are spatially or temporally uniform; yet in early applied hydrology predictions tended to be based on an assumption that all the catchment contributes overland flow down straight, parallel slopes to a homogeneous receiving channel. The obvious question is, are there regularities within the irregularities which we now acknowledge? Fortunately, as the Horton model of overland flow has become upset by field observations it has been possible to adopt two new ways of envisaging both quickflow and baseflow generation. The first of these demarcates a relatively small part of the catchment area which remains saturated or very moist—the 'partial contributing area' (PCA). The PCA tends to be located on valley floors, in wetlands, and in hillslope hollows—we have already covered the hillslope processes responsible in Section 3.5.3.

Mapping of runoff frequency in the field has established that key areas of the catchment yield overland flow with greater or lesser frequency (see Fig. 3.13). The partial contributing area might be considered as that yielding quickflow (rather than overland flow) but there are clearly complications thanks to the phasing in and out of different flow routes during an event. Nevertheless, using saturation indications, such as natural vegetation, the PCA can be delimited without the huge field effort, for example of Weyman (1975), and simple rainfall/runoff predictions adjusted to the PCA rather than the whole catchment. The remainder of the catchment plays a part, clearly, in producing the saturated areas and other parts of the PCA, often between storms or snowmelt events.

3.7.3 Dynamic contributing areas

The variability of flow routing within segments of slope, the extension of the surface-channel network,

and the operation of valley-floor processes during a sustained flood event all conspire to render the PCA concept of limited reality, particularly in headwater catchments. Here one may observe the PCA extending as saturated soils and surface flows respond to the continuing precipitation inputs. As the hydrograph of flow builds to a peak and then falls to its recession we can record areas contributing quickflow as waxing and waning too. Figure 3.14

Table 3.5. Slope–channel linkages in the humid-temperate river basin

Headwater
As in GORGE in Table 3.4: direct contribution of runoff and slope material if available. Both are likely to be channelled into main streams by the high density of tributaries. Rapid response by both runoff and sediments.

Valley floor
In glaciated terrain, valley floor is widened by glaciation and infilled by subsequent slope processes from over-steepened valley sides. Valley floor often composed of impermeable drifts and therefore both saturated, carrying overland flows or return surface flows, and unstable at the channel margin-forming river cliffs.

Floodplain
Contributing slopes are now distant and exercise little influence on quantity/quality of runoff, both of which are modulated by the floodplain's storages. Channel routing of flows, sediments, and solutes becomes an important control on river management.

Fig. 3.13. Frequency of surface runoff in a small headwater catchment (after Weyman, 1975).

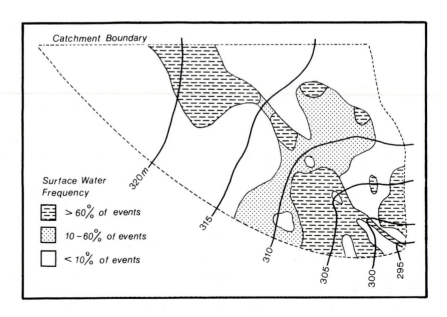

Catchment Boundary

Surface Water Frequency

> 60% of events

10 - 60% of events

< 10% of events

320 m
315
310
305
300
295

adds the waxing and waning of the surface-channel network to this picture, highly simplified for clearer presentation.

If we return to the problem of process modelling and retain our conceptual framework of the runoff 'cascade' as a series of overflowing tanks the 'dynamic contributing area' (DCA) principle involves a profound modification to the slope-channel links which have to be made state-dependent, i.e. made to vary according to the state of storage and flow in adjacent tanks (see Fig. 3.1).

Plate 3.5. (a) Soil pipe discharging return surface flow at blockage during heavy rainfall. (b) Network of desiccation cracks through which the pipe is developed (centre).

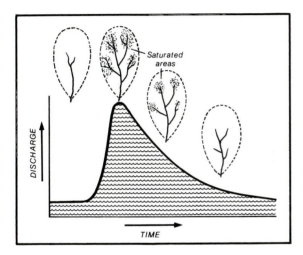

Fig. 3.14. The relationship in space and time between surface runoff and the contributing area of a catchment.

3.7.4 Pipes and seeps

Putting our saturation-based DCA principles into a slope-channel section diagram, Fig. 3.15(a) reveals a considerable complication—the presence in some soils of soil pipes (and in others of lines of rapid seepage through macropores). According to a comprehensive review by Jones (1981), soil piping can occur under the following conditions: (a) desiccation cracking of soils; (b) mass-movement cracking of soils; (c) tunnelling in dispersive soils (i.e. those lacking chemical bonding).

They are clearly, therefore, influential in conducting quickflow (in the case of ephemeral pipes) or baseflow (in the case of perennial pipes) very efficiently from source areas to the nearest surface channel. Fig. 3.15(b) reveals how ephemeral and perennial pipe networks link together on a hillside in Wales. It is, however, erroneous to consider these

networks always as extensions to, and in perfect continuity with the permanent open-channel network. Pipes reach river banks relatively rarely. At a certain stage of flood-flow generation they can be important (Plate 3.5(a)) but they have little available capacity and if they do not spill as return surface flow at the edge of the valley floor they become blocked or collapse upslope, forcing runoff into the hydraulically rough vegetation zone, effectively slowing it. The origin of the ephemeral pipe networks in British upland soils appears to be erosion of desiccated peat (Plate 3.5(b)).

3.7.5 Scale effects in the river basin

Considering a simple, threefold classification of slope runoff into Hortonian, saturation overland flow, and throughflow opens up the possibility of synthesis of

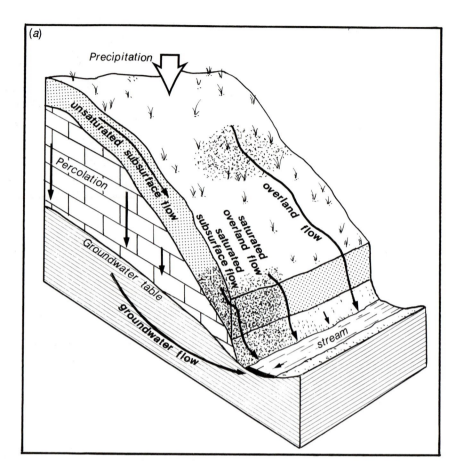

the volume and timing of catchment runoff at various scales. The regularities of stream ordering and drainage density are also inputs to this assembly of processes: basically the time spent on slopes dominates the performance of a water droplet in headwaters, but lower down the basin the time of travel in channels comes to equal and exceed that on slopes.

Nevertheless, such syntheses are fraught with difficulties: increasing scale often crosses climatic (notably rainfall) boundaries and soil types often show zonation away from headwater areas. Whilst statistical prediction in hydrology often lumps these complicating variations within the use of drainage basin areas as a variable, Pilgrim (1983) warns of the dangers of transferring process-based findings from small to large basins and between regions. The problems still confront hydrological modellers, who are forced in many cases to model the outflow hydrograph for a series of small headwaters or *routing* to 'get the water downstream' to larger basin outlets.

3.8 Runoff processes outside the humid-temperate zone

Perhaps the most obvious reason for variation in the processes of runoff between climatic zones is difference in the form of precipitation received by the river basin concerned. Thus regions of ice and snow conditions, either of annual or seasonal duration, are likely to show fundamentally different processes from those in rainfall zones; not only is the phase of the incoming precipitation different, but the ground conditions are radically different from those with vegetation and soil covers. In terms of process models we may consider this equivalent to the addition of another 'tank' at the top of the runoff cascade.

Snowmelt is highly important to a number of runoff regimes in the world, not only to those where nival floods occur (Section 4.6) but notably to those semi-arid areas naturally or artifically irrigated from neighbouring snow-capped mountain ranges (e.g. Colorado, USA, and parts of Iran).

It will form a useful summary of *snowmelt-runoff*

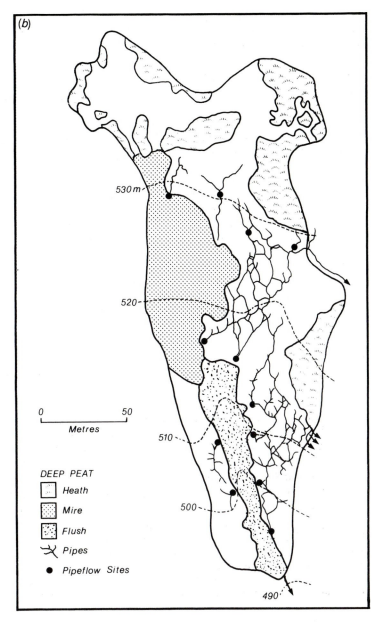

Fig. 3.15. Hillslope runoff: process and pattern. (a) Flow routes on hillslopes. (b) Distribution of soil pipes on a small section of hillside in mid-Wales, UK (after Gilman and Newson, 1980).

(b)

530 m

520

0 — 50
Metres

510

500

DEEP PEAT

Heath

Mire

Flush

Pipes

● Pipeflow Sites

490

processes if we consider briefly the main techniques used to predict the runoff rate and volume in nival river basins (see Morris, 1985). The simplest techniques involve calculations of the mass of snow or ice in the catchment pack (from field survey or satellite imagery) and the energy input needed to bring the pack to melting conditions. Air temperature, in degree-days over a melting threshold, is normally used as a simple surrogate for the group of energy routing processes involved with snowmelt. This energy cascade involves the albedo of the snowpack, the turbulent transfer of energy from winds crossing the pack, conduction and convection of energy from the soil beneath the pack, and the energy transmitted to the pack by infiltrating melt (the pack acts as a surface soil horizon). A problem also arises under conditions of 'rain-on-snow' where the water equivalent of the pack increases with the rainfall but energy (influencing melt rates) is also transmitted via the infiltrating rainfall. It is not necessary to model the entire suite of processes to obtain an acceptable prediction as Ferguson (1984) proved with a simple model based upon height/temperature and area/depth curves for a small catchment in the Scottish Highlands.

Further indications as to the nature of snowmelt processes can be obtained from the way in which they become altered by land-use changes such as urbanization. Buttle (1990) has shown how, in a small urbanizing catchment in Ontario quickflow was increased fourfold during snowmelt events. This was not the corollary of the spread of impervious surfaces, for snow is often cleared in cities to piles on permeable areas such as gardens. Other factors ensue, such as the radiative-energy input from housing and the reduced albedo of 'dirty' piled-up snow packs.

Forests also produce a contrast in runoff under snowy climates; they both 'scour' snow efficiently and therefore have more deposited than in surrounding clearings and open country (Shiklomanov and Krestovsky, 1988), but also evaporate it very efficiently under certain conditions (mild winds) from an interception store (see Calder, 1990) and also protect it from melt if it penetrates to the shade beneath the canopy, where it will melt only slowly in avalanched 'drifts'.

3.8.1. Runoff processes in semi-arid areas

The precipitation regime of semi-arid areas is extraordinarily variable; in this regime, therefore, runoff processes reflect a combination of the rarity and intensity of the rainfall itself and the condition of the surface under hot, dry climates.

Yair and Lavee (1985) have summed up the situation as follows,

The scarcity of storms, the limited rainfall amounts, the shallow and patchy soils, and the low-density vegetal cover, inhibit significant contributions to storm channel runoff by groundwater flow, saturated overland flow or shallow subsurface flow. As infiltration capacities are lower than the rainfall intensities of most storms, overland flow could therefore be considered to be high in frequency and magnitude. (p. 185)

Thanks to the prevalence of intensive convective rainfall as the origin of runoff in such areas it has required radar and satellite rainfall assessments to compile co-ordinated rainfall-runoff observations (Section 2.10 refers to the hydrometric problems). Rainfall intensities reach 90 mm/hr (compare with the newsworthy fall of 20 mm in an hour in southern England as this book goes to press!). Field measurements have come from Arizona, Israel, and southern Spain. For example, Reid and Frostick (1987) derive a time-to-peak duration of 4–16 minutes for a 7 km^2 catchment in Kenya; it has a drainage density of wadis of 100 km km^{-2} and hence the rapid development of ponded surface water and overland flow leads to very steep hydrographs. In fact a shallow 'bore' of water travels down the wadis; transmission losses to the alluvial fills are very large. This perhaps explains why, despite their very short rise-times in floods their peak flows are similar to those of the humid temperate zone—swings and roundabouts!

Both Yair and Lavee (1985) and Pilgrim et al. (1988) list the runoff differences between the semi-arid zone and humid zone, cataloguing the variability of semi-arid land surface from which runoff occurs; the shortage of soil and vegetation cover sets up considerable contrasts between bare rock and weathered rock, for example. Reid and Frostick confirm this variability with infiltration rates of 78 mm/hr on sandy soils (10 mm hr^{-1} when they are crusted after long dry periods) and next to no infiltration on bedrock.

3.8.2 Runoff processes under tropical forests

Catchment studies from the humid tropics are, for obvious reasons of access and definition, much rarer

---

Box 3.2. *Some examples of high intensity rainfall from tropical regions*

Duration	Rainfall (mm)	Location	Date
15 min.	40.6	Ibadan, Nigeria	16.6.1972
90 min.	254.0	Colombo, Sri Lanka	(1907)
24 hr.	1168.1	Baguio, Philippines	14.7.1911
48 hr.	1671.0	Funkiko, Formosa	19.7.1913
96 hr.	2586.8	Cherrapunji, India	12.6.1876

in the hydrological literature than plot studies. The popularity of the latter scale of working also arises from the dire need for studies of deforestation effects on runoff processes, erosion rates and loss of nutrients. Ross *et al.* (1990) report on experiments in the Amazon rainforest in Maraca, Brazil. They reveal a domination of the natural runoff processes by throughflow, which in turn is influenced by the location in the soil profile of structural layers such as laterites and, importantly for the study of felling, by location on the slope profile. Leaf litter and soil organic matter are also critical to the encouragement of infiltration beneath the forest cover under conditions of extreme rainfall intensities (Box 3.2). Mahoo (1989) reports a very comprehensive study in south-east Nigeria in which a range of treatments (both felling and cropping) was applied to the tropical forest; detailed studies of runoff processes, sediment yield, and runoff quality were conducted with the following main results:

- Runoff is dominated by throughflow under all conditions.
- After clearing, especially mechanical clearing, the proportion of surface flow increases.
- The reasons for this change include an increase in bulk density of surface soils, a decrease in porosity, and hydraulic conductivity.
- Erosion rates and nutrient losses increased above the very low rates in the forest control plots.
- The nature of subsequent cropping and the treatment of those crops had additional effects.

Despite these useful and far-reaching studies the volume of hydrological knowledge emerging from this critical zone for the planet's hydrological cycle remains tiny compared with the magnitude of the problem of sustainable management, effectively a problem of hydrological management of vegetation and soils.

4. Runoff extremes

4.1 Extremes in principle

The history and philosophy of the natural sciences are full of references to the role of extremes in the pattern of major forces, geological and climatic. Early interpretations were based on religious belief in the wrath of a creator-deity. Science then contributed concepts of timeless activity of global processes in which extremes were normal and could be quantified, using statistical analyses, with a view to predicting their impact and hence protecting against them. Recently, however, there has been something of a trend back towards *catastrophism* (Huggett, 1989), i.e. a special role for extremes in, for example, bringing about geomorphological change (see the effectiveness of extreme floods considered by authors in the volume edited by Beven and Carling, 1989).

In this chapter we deal with the hydrological extremes of flood and drought, the collection of data on them, its analysis, the predictive role of such analyses, and the reaction of human and ecological systems to both the extremes and protection against them. We need to do this under a general theory of natural and man-made hazards; politicians and planners remain confused by hazards, yet many researchers, notably geographers, have laid down principles which, if acted upon, would yield a much more sustainable approach to protecting communities, (e.g. Hewitt and Burton, 1971). The essential message of this approach is that:

$$\text{RISK} = \text{VULNERABILITY} + \text{HAZARD}$$

an equation in which all terms are dynamic. Communities regularly increase their *vulnerability*, either in search of other resource gains or because they are poor and have no choice. *Hazards* vary in

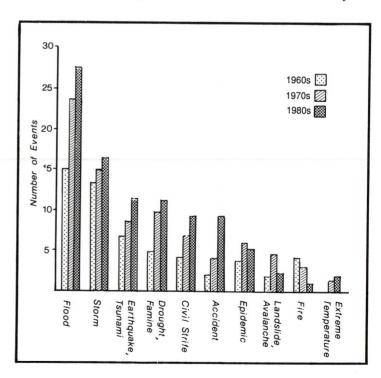

Fig. 4.1. Average numbers of disaster events, worldwide, of various categories. (Data for 1980s restricted to 1980–1.) (After Wijkman and Timberlake, 1984.)

Table 4.1. The changing impact of hazards on human communities and the relative roles of drought and flood

Type of event	People affected		Lives lost	
	1960s	1970s	1960s	1970s
Drought	18 500 000	24 400 000	1 010	23 110
Flood	5 200 000	15 400 000	2 370	4 680
Civil strife/ conflict	1 100 000	4 000 000	300	28 840
Tropical cyclone	2 500 000	2 800 000	10 750	34 360
Earthquake	200 000	1 200 000	5 250	38 970
Other disasters	200 000	500 000	2 890	12 960
TOTALS	27 700 000	48 300 000	22 570	142 920

Source: Wijkman and Timberlake (1984).

location and intensity as part of environmental change but both these characteristics are also altered by development processes.

We need to put floods and droughts into a general hazard context in terms of occurrence (in space and time), fatalities, damage costs, and other statistics. The reviews by Wijkman and Timberlake (1984) and by UNEP (1991) are a basic source of reference here. Fig. 4.1 illustrates the incidence of floods and droughts in relation to other disaster events in the last three decades. Whilst numbers of events appear to be rising this is perhaps a result of better reporting; however, as Table 4.1 reveals, floods and droughts also appear to be affecting more people (especially poor people in the developing world).

As reported by the World Commission on Environment and Development (1986):

The disasters most directly associated with environment/development mismanagement—droughts and floods—affected the most people and increased most sharply in terms of numbers affected. (p. 7)

4.2 Extremes: mankind and other biota

It is easy to concede that *Homo sapiens* has a larger investment in steady-state conditions of the natural elements than other species: modern mankind has a settled existence, extracting localized and specialized services from the environment of shelter, food, and energy supply. Adaptability to extreme conditions has largely been sacrificed in development to a notion of comfort and welfare; mankind also shows charitable concern for those on whose lives elevated extremes have wrought destruction and havoc:

earthquake, drought, and flood. Resources therefore become transferred to hazardous locations and specialist human skills, such as engineering, are deployed to provide protection. As general texts on natural hazards reveal (e.g. Kates, 1978; Brown, 1989) such protection often merely increases the risk of disastrous human effects for natural extremes. Furthermore, protection for one part of a system may increase risk for another, as when flood protection for an upstream city exacerbates flooding downstream because the flood waters are denied storage on the floodplain.

Whilst this chapter takes a largely anthropocentric view of hydrological extremes (their prediction has formed a major, central role for hydrology) it is worth considering that the effects of extremes are different for other biota.

There are four main responses by living organisms to extremes in the immediate environment (see Fig. 4.2):

Absorb the 'shock'
Accept the loss
Reduce the impact
Change habits or move.

These reduce to two main categories of coping—*adaptation* and *adjustment*. Humans have more technological options for both strategies than other biota, which may adapt on evolutionary time-scales but whose most frequent immediate adjustment is to move. For example, a major flood accompanied by peat erosion which affected the River Tees in 1983 killed an estimated 7600 fish in an 18 km reach (McCahon *et al.*, 1987)—largely through the impact of suspended sediments and dissolved metals—but invertebrates were less affected and fish stocks have recovered by natural restocking from tributaries and from downstream where the direct impacts of the flood diminished with increasing catchment area.

Pollution episodes may also be regarded as environmental extremes to biota. Freedman (1989) lists eighteen ways in which ecosystems respond to stress of this type. The four main characteristics are adjustments of energy cycling, nutrient cycling, community structure and linkages between organisms.

4.3 'Threshold' reactions

Environmental scientists continue their quest for a general model of extremes and their effect on

Fig. 4.2. Models of coping with natural hazards (after Burton *et al.*, 1978).

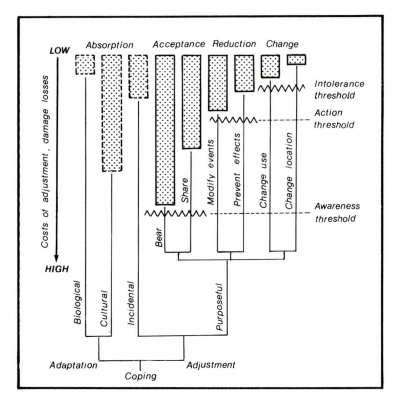

physical variables. A successful conclusion would allow us better insights into the significance of extremes (including the degree of independence between sequential events) and a better judgement of how to respond in various forms of coping.

The concepts of *thresholds* and metastability are becoming popular as interim models for sudden and rapid change in natural systems. As Fig. 4.3 indicates, the dynamic equilibrium state of a system, such as a river channel, may be disturbed in the direction of the extremes of its range with either temporary or permanent effects. In the former case there is 'recovery' of the system, so that the river channel, which may have had bankside vegetation scoured by a flood, revegetates with the same species. The new equilibrium may be slightly different if, for example, bank erosion is such that trees will never find suitable conditions for growth again as they did pre-flood. Quasi-permanent threshold changes modify the main system variables, preventing recovery—as might be the case where flood erosion of the bank was accompanied by a landslide which transforms the channel into a sandy bed when previously it was composed of boulders; channel shape will adjust to the new substrate as will stream and bank plants and animals.

There is still much conjecture on the topic of thresholds but the insights which the concept provides are of practical use in designing mitigation devices for a disaster which, in the longer term, may be more apparent than real. In reverse, if a system has crossed an irreversible threshold those exploiting it must adjust to its new controls whether they are riverside factories or communities of invertebrate animals or algae.

It is now of urgent research interest in hydrology to assess the probable types of change during the changes of climate likely to accompany the enhanced 'greenhouse effect' (Newson and Lewin, 1991) and how they relate to the patterns of change resulting from man-made activities such as those described in Chapter 5.

4.4 Floods and the flood hazard

The gathering of comparative statistics on natural hazards has a short history and systems of data collection are least well organized in some of the most hazardous zones; in the case of flooding, national and international statistics tend to lump together flooding from both river and sea. One of the charac-

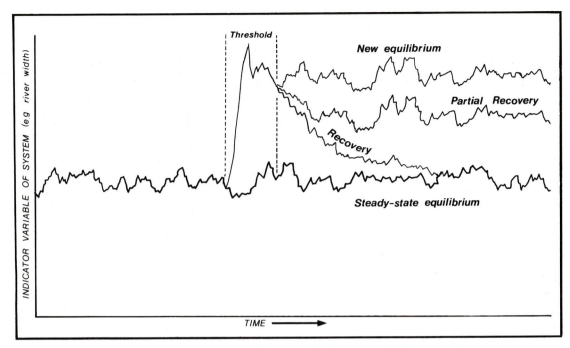

Fig. 4.3. Pictorial representation of states in a natural system vulnerable to threshold change behaviour.

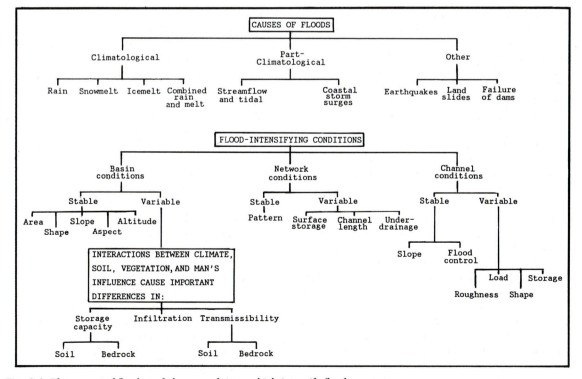

Fig. 4.4. The causes of floods and those conditions which intensify flood events.

teristics of the flood hazard is its multiple origins and we develop this element of naïve unpredictability in Fig. 4.4 which also indicates 'flood intensifying conditions'. These can prove so sensitive in certain floods that neighbouring basins react very differently because of their size, physiography, land-use or because of the rainfall or snowmelt characteristics of the event. Clearly, too, dam-burst flooding is most unusual as a hazard since it results from protecting (inadequately in terms of construction!) against drought.

4.4.1 Measurement, data problems

The problems of flood-flow measurement are well known to all field hydrologists; they include considerable personal danger! The characteristics of a river channel listed in Table 2.7 as part of selecting an appropriate gauging technique are fundamentally altered during an extreme flood, in ways listed by Table 4.2. As a result, techniques relying on structures fail because they are bypassed or damaged; velocity measurements during the event are prohibitively dangerous and even gauge-boards are washed away. For these reasons, the collection of accurate flood peak-flow data, upon which statistical analysis relies, often needs to fall back on detective work. Velocities may need to be reconstructed from the strand line (*trash marks*) left by lighter debris at the margins of the flow; the slope of such lines can be entered in the Manning equation (Section 2.5.1) and trash marks can also be used to indicate the cross-section of the flow at maximum flood stage. Other strategies may be employed, including those to 'rescue' rainfall data from over-spilling rain-gauges in the flooded catchment, as shown in Table 4.3.

Flooding in *arid and semi-arid catchments* poses special problems of gauging (Rodier and Roche, 1978). Channel definition is a problem in landscapes lacking vegetation; channels also scour and fill rapidly during successive floods, undermining or bypassing gauging equipment. Flood flows may be propagated as 'bores' down the valley, velocities of flow are extremely high (as are sediment loads); these factors combine to militate against the use of current meters and other gauging techniques. It is often overlooked in the comfort of a humid climate that transport to gauging sites may be 'off road' and very hazardous during flood flows. In such environments an understanding and quantification of flood flows can be expanded by use of geomorphological reconstruction techniques for peaks (Richards *et al.*, 1987).

A very useful source of river peak flow data is the work of local people in marking the top water levels with a plaque or paint mark etched with the date of the event (Plate 4.1). If these *historic flood marks* are surveyed in relation to the river channel, and assumptions are made about velocities, flood series can be extended with great benefits for statistical prediction (Leese, 1973; McEwen, 1989).

However, care must be taken that, in the case of historic flood marks, the channel has not changed its capacity or been modified by dredging since the flood. Over periods of a century or so we can expect quite profound changes in channel capacity, spoiling our chances of reconstruction (Macklin *et al.*, 1992).

Geomorphologists can help estimate peak flood-

Table 4.2. Changes in river-flow characteristics which threaten the measurement of flood flows

Characteristic	Normal	Flood
Depth/width	Within banks or within structure.	In floodplain sites station is bypassed. Floodplain flows difficult to measure. Flow structures damaged by force of deep water.
Velocity	Measurable by instrumental method wielded by field observer.	Life-threateningly high, even in small streams. Pattern of isovels disrupted. Instruments damaged. Control on velocity passes downstream or to temporary obstructions.
Solids transport	Structures not normally used on rivers with high solid load.	Flood sediment transport fills structures and stilling wells. Shoals control water levels. Instruments damaged by impact; vegetation 'snags' instruments.

Table 4.3. Strategies employed to rescue data from important UK floods

Flood, date	Data recovery techniques	Reference
Bruton, 1917	Rain-gauge inspection/calibration.	British Rainfall (1917)
Lynmouth, 1952	Use of trash marks and largest material moved on river bed.	Dobbie and Wolf (1953)
Camelford, 1957	Assessment of hail lost to raingauge. Boulders moved.	Bleasdale (1957)
Exeter, 1960	Use of photography of flow patterns around obstacles such as lamp-posts.	Harrison (1961)
Mendips, 1968	Flow levels in cave passages.	Hanwell and Newson (1970)
Plynlimon, (1973)	Use of trash marks + Manning formula.	Newson (1980)
Cairngorms, 1978	Use of channel change/sediment size/trash marks	McEwen and Werritty (1988)
Howgill Fells, 1982	Morphology of hillside scars and deposit.	Wells and Harvey (1987)
County Leitrim, 1986	Peat-slide morphology and boulder berms.	Coxon, Coxon and Thorn (1989)
Bradford, 1989	Use of bucket surveys, indications from rainfall radar	Acreman (1989)

Plate 4.1. Flood marks on the River Severn, UK.

flows from the *location and calibre of coarse flood deposits*; however, complications exist where landsliding as well as channel transport has figured in the event. Landslides also damage rain-gauge networks; rainfall reconstruction or rescue techniques are also invaluable in the aftermath of severe floods. If gauges are not damaged they may overflow because they have not been reached to empty or may be reached by flood waters and 'drowned'. A 'bucket survey' may be needed of any container left in the open during the event whose volume of catch may be converted into a depth. Altitude may be the basis of extrapolation of these depths to the whole catchment; clearly rainfall radar records obviate the need for these extremely tentative reconstructions (see Acreman, 1989).

4.4.2 Data series, abstracting data

Assuming for the moment that a long and continuous record of flood flows exists for a catchment (or that 'missed' peaks have been reconstructed by successful detective work) what information can be gleaned which will help our understanding of flood processes and to predict or forecast future occurrences?

Clearly, information is available from recording gauges for both *instantaneous flood peaks* and for the entire *flood hydrograph* of each event. The former will indicate the comparative extremes of each event in terms of river level (and hence, perhaps, velocity and width as well) whilst the latter, when pooled and averaged, will indicate duration of flooding, rise times and other important aspects. The flood hydrograph also provides a revealing link with catchment processes (and artificial influences on them) as revealed in Chapter 3.

Fig. 4.5 illustrates these two channels of flood information in a long flow record but also indicates a dilemma in selecting flood-peak data. What is the relevance of, for example, a secondary peak 'riding on' the hydrograph of the highest discharge in an annual period? Should merely the *annual series* of highest peaks each year be used on statistical analysis (yielding, clearly, a probability of occurrence per calendar year), or is the calendar year an unnatural abstraction? Instead, should use be made of all flood peaks in a record to obtain maximum information? The compromise solution in many flood studies is to use 'peaks over a threshold' (Fig. 4.5, upper) thus forming a data set known as the *partial duration series* to distinguish it from the annual series. (Please note that this use of the term 'threshold' is purely numerical or procedural and is not connected conceptually with the threshold behaviour of systems discussed in Section 4.3.) Further refinement may be employed to fix the threshold such that the number of peaks selected corresponds

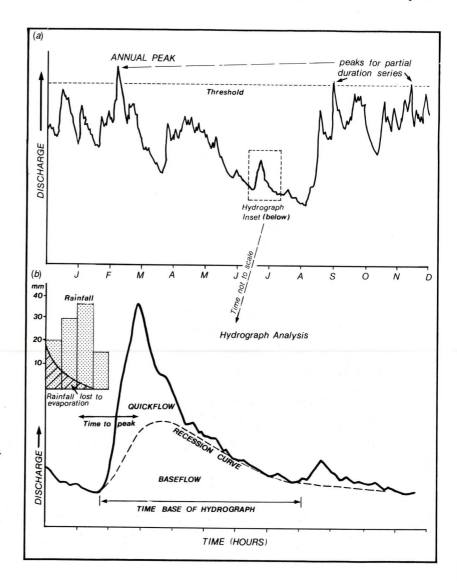

Fig. 4.5. Data types for flood analysis. (a) An annual time series of data from which the peak(s) of discharge is (are) selected for probability analysis; (b) The hydrograph and causative rainfall for an individual flood. These are compared by unit hydrograph analysis.

Box 4.1. *Fitting a gumbel plotting position to an 84–year flow record*

Year of record	Annual peak flow (cumecs)	Rank	Year of record	Annual peak flow (cumecs)	Rank	Year of record	Annual peak flow (cumecs)	Rank
1	292	46	30	367	28	58	410	15
2	231	65	31	255	55	59	384	19
3	230	67	32	256	54	60	298	44
4	244	59	33	585	3[c]	61	457	11
5	284	48	34	373	24	62	115	83
6	208	73	35	327	34	63	261	52
7	237	61	36	351	30	64	257	53
8	205	74	37	334	33	65	714	2[d]
9	171	81	38	251	56	66	227	70
10	339	32	39	240	60	67	299	43
11	300	42	40	198	78	68	324	35
12	173	79	41	231	64	69	385	18
13	789	1[a]	42	298	45	70	377	20
14	202	76[b]	43	522	7	71	263	50
15	351	29	44	370	25	72	231	64
16	171	80	45	375	23	73	453	13
17	262	51	46	526	6	74	316	38
18	533	5	47	235	62	75	314	39
19	200	77	48	552	4	76	317	37
20	162	82	49	228	69	77	375	22
21	386	17	50	274	49	78	308	41
22	516	8	51	478	9	79	456	12
23	229	68	52	95	85	80	344	31
24	249	57	53	227	71	81	286	47
25	220	72	54	478	10	82	369	27
26	376	21	55	438	14	83	113	85
27	204	75	56	247	58	84	324	36
28	231	66	57	369	26	85	313	40
29	395	16						

$$\text{Recurrence interval in years} = \frac{\text{years in record} + 1}{\text{rank of flood}}$$

Examples : [a]86-year flood; [b]1.13-year flood; [c]28.6-year flood; [d]43-year flood.

Source: Modified from Wilson (1983).

with the number of years of record (Shaw, 1983); the data series is then labelled the *annual exceedance series*. Each data series is appropriate to both a particular flood regime and a particular length of record (e.g. partial duration for short records) but each makes assumptions of statistical independence which must be matched with the appropriate probabilistic treatment. The reader must follow the detail of statistical arguments to be able to work outside the straightforward instruction of flood manuals (e.g. NERC, 1975).

4.4.3 Probabilistic treatment of peaks

Whatever data series is compiled for a gauged catchment the hydrologist charged with aiding the response of a community in terms of flood protection will wish to express the likelihood of a flood of a particular discharge recurring at the gauged site (or at a nearby gauged site with the results carefully transferred—for ungauged sites see below). This approach is the equivalent, in a sophisticated, formal way, of fixing betting odds on the basis of a racehorse's form. On average, how often does a flood of x cumecs occur?

In order to answer this question we need to know from statisticians which *statistical distribution* fits the floods-data series we have; we will have very little information in the very zone of the distribution where we need most—by definition there are not many rare floods for us to record! Therefore, the chosen statistical technique must perform well in the 'tail' of the distribution; we may choose to transform the flood peaks into their logarithms before fitting the distribution to improve this performance.

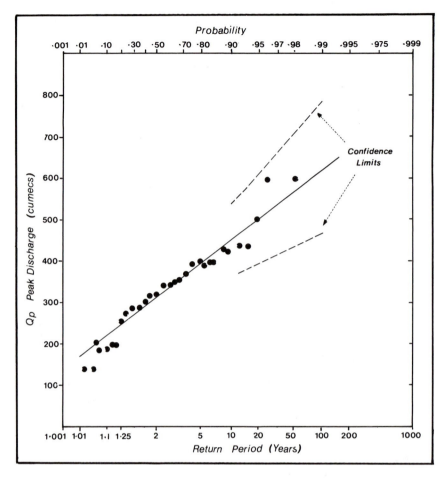

Fig. 4.6. Analysis of return period (recurrence interval) of annual floods using probability plots in the analysis pioneered by Gumbel.

The most persuasive distribution used in the hydrology of extremes is the *Gumbel distribution* (or Generalized Extreme Value distribution); it is both easy to calculate how to plot each flood peak (Box 4.1) and the appropriate probability graph paper is widely published (Fig. 4.6). Nevertheless, rivals appear regularly and Wilson (1983) illustrates the sensitivity of this choice in terms of the effect on calculating very large floods.

Knowledge of this sensitivity of chosen distribution and the general shortage of rare flood data encourages hydrologists to pool data within hydrological regions to predict the peaks which occur, on average, every 100+ years. Records at a site will generally prove adequate to calculate the *mean annual flood* or the 5-year, 10-year floods (used, for example in designing road culverts or farm-drainage schemes). The ratio of rarer floods to the mean annual flood in catchments where these higher floods are gauged can then be plotted regionally (Fig. 4.7). These curves not only give more

confidence about the return period of major floods where there exists a short gauging record but, since the mean annual flood can be related to the physiographic and climatic *characteristics of the catchment* by regression analysis (Table 4.4), we have a technique for predicting floods on *ungauged catchments*.

Regional regression analysis of the mean annual flood on catchment characteristics has been carried out for many regions of the world and Table 4.5 is by no means an exhaustive compilation. Fig. 4.8 combines a guide to the probabilistic use of flood-peak data which we have just described as well as the alternative use of unit hydrographs and a probabilistic approach to rainfall.

4.4.4 Separating probability from risk

In time, with refined statistical techniques and very long data-sets, and if climate remains stable, we may be equipped with very precise methods of flood prediction. However, we have already concluded

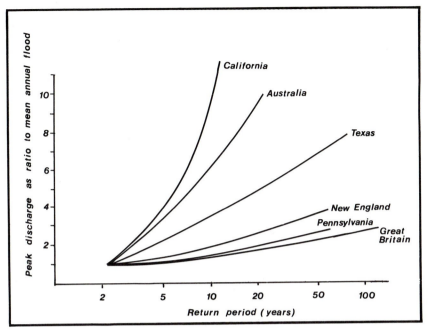

Fig. 4.7. A comparison of regional growth curves for floods of high return periods (data from Institute of Hydrology, Wallingford, UK).

Table 4.4. Correlation matrix of Mean Annual Flood (MAF) and simple catchment indices

Rainfall	MAF	Area	Slope	Urban	Lake
MAF	0.75	−0.18	−0.08	0.08	0.40
Area		−0.68	−0.09	0.11	−0.12
Slope			0.01	−0.10	0.51
Urban				0.09	0.21
Lake					0.25

Note: In the form of a prediction equation the variables appear as MAF $= 0.0201$, $AREA^{0.94}$, $STMFRQ^{0.27}$, $SLOPE^{0.23}$, $SOIL^{1.23}$, $RSMD^{1.03}$, $(1 + LAKE)^{-0.85}$; STMFRQ and RSMD are defined in the source reference.

Source: NERC (1975).

Table 4.5. Successful catchment variables in flood prediction

Authors variables	Flood parameter	Successful
Benson (1962)	Mean annual flood	A, S, Ri, St
Carlston (1963)	Mean annual flood	A, Dd
Gray (1964)	Lag time	L, S
Gupta *et al.* (1969)	Peak discharge	A, L, Lca, S
Nash (1960)	Hydrograph indices	Sc, L
Rodda (1967)	Mean annual flood	A, Dd, Ri
NERC (1975)	Mean annual flood	A, S, U, St, R, C

Key: A = Catchment area; S = Main channel slope; Sc = Catchment slope; Ri = Rainfall intensity; Dd = Drainage density; L = Mainstream length; Lca = Length of stream to centroid of area; St =Area of storage in lakes/reservoirs; U = Urban area as proportion of total; R = Rainfall (av.); C = Climate measure.

that floods have many basic causes and many intensifying features. In climate terms there may be a flood regime for any given catchment which includes snowmelt floods, convection-storm floods, and frontal floods. Recently Hirschboeck (1988) has called for a *hydroclimatology of floods* (of special relevance under conditions of changing climate) and Fig. 4.9 shows how probabilities may differ for flood peaks of differing climatic origins. Figure 4.7 also indicates a regional climate impact on growth curves for extreme flood peaks.

Probabilities, however sophisticated, need translating into *risk* (both in statistical and perceptual terms) before flood designs are attempted. The difference between probability and risk in statistical terms can be exemplified by a channel enlargement scheme designed to contain the 100-year flood. To build it, we need to put heavy machinery in the river; what is the risk of the 100-year flood (or any other magnitude of event) occurring during the year we are constructing the scheme? Because we are protecting settlements against events whose occurrence we cannot forecast in real time the protection schemes will have a *design lifetime* which is shorter than the recurrence interval of the design flood. For example, there is a 33 per cent risk of a 50-year flood occurring within 20 years of a scheme designed to protect against it (it may also happen tomorrow!). Wilson (1983) tabulates the full relationship between design lifetime and recurrence interval.

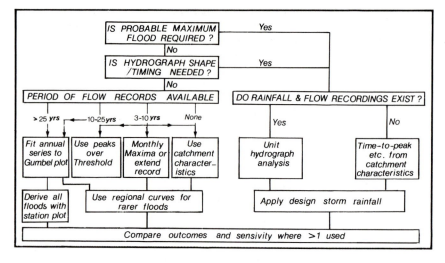

Fig. 4.8. A simple guide to the selection of a flood analysis/predictive technique for engineering design in the British Isles.

The other aspect of linking probability to risk is the choice, for any particular job of flood protection, of a design probability (inferring a risk of exceedance and therefore a cost to the affected townspeople, travellers, farmers, etc.). Some commonly chosen values are:

- Floodplain grazing land — 3 year intervals
- Floodplain arable land — 10–25 year intervals
- Urban centres — 50–100 year intervals
- Reservoirs in remote areas — 150–1000 year intervals
- Reservoirs in populated areas — 10 000 year intervals

but clearly this choice is more within the realms of social science. Economists will be asked to relate the cost of successively more ambitious levels of protection to reduce the risk of flooding against the financial benefits of so doing; where the *ratio of benefits to costs* exceeds unity the scheme is viable. For this approach to succeed we need another kind of flood data—that on damage costs. The depth–damage curve approach is illustrated in Fig. 4.10. Composite curves for neighbourhoods can be agglomerated from sample surveys (Penning-Rowsell and Chatterton, 1977).

4.4.5 Hydrograph analysis/synthesis

The hydrograph of river-flow response contains a wealth of information about the mechanisms and routes by which a rainfall signal is translated into a flood response: a pattern bespeaks a range of processes as we observed in Chapter 3 (e.g. Fig.

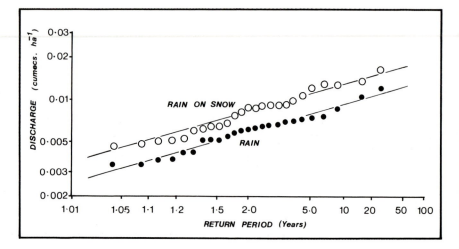

Fig. 4.9. Partial duration series plots for flood peaks from different climatic origins, Oregon, USA (after Church, 1988).

3.2). However, disentangling the information is difficult because of the habit of natural processes in smoothing and overlapping. Furthermore, in order to carry out simple and robust analyses of the pattern of flow response we need a relatively gross simplification, e.g. a triangle (see below) to work with. Analysis then runs the severe risk of ignoring the basic process information in the hydrograph such as rates of loss by evaporation during the flood (do not assume zero), the relative volumes of surface and subsurface flows, or even the direction of movement of the rain-storm which caused the flood.

The first model of the hydrograph developed for engineering use was the *unit hydrograph* (Sherman, 1932), so called because the analysis detects the pattern of response to units of precipitation input— allowing comparison between events and application via known rainfall information for an *ungauged catchment*. Indeed there have been two major driving forces behind the refinement of hydrograph techniques:

- the generally wider availability of lengthy *rainfall records* than flow records;
- the fact that sites such as dams which require engineering designs for *extreme flood* protection do not often have any previous flow gauging.

Successful hydrograph techniques can also be used for flood-flow simulation under very extreme rainfall or other precipitation conditions (see Section 4.4.6). Despite the success of unit hydrograph techniques since the 1930s hydrologists have been mindful that the needs of application have often driven new empirical developments of the basic analyses; there have been frequent calls for more sophisticated mathematical treatments (e.g. authors in Anderson and Burt, 1985) or for theoretical advances in the hydrology of catchments to be incorporated in *hydrological models* focusing on flood response.

Retaining the traditional unit hydrograph technique here, analysis may be carried out in very simple ways such as described by Wilson (1983)—a matrix of rows representing time periods suitable for

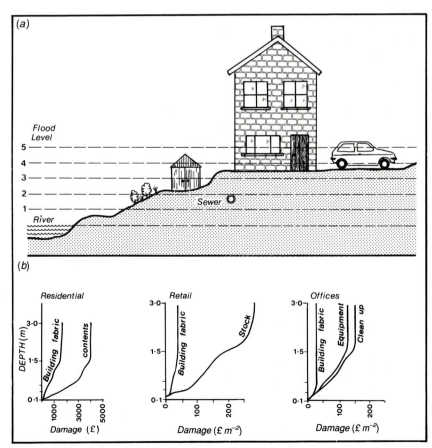

Fig. 4.10. Flood depth/damage. (a) Relationship between river level and property damage of different types. (b) Quantified depth/damage curves for various premises (after White, 1964).

the catchment (e.g. one hour) and columns representing rainfall, losses by evaporation, and the baseflow estimate (which is subtracted from the total flow). This approach is suitable for spreadsheet analysis on computers whilst more sophisticated dedicated hydrograph programs are also widely available.

As in the case of the probabilistic treatment of flood peaks, grouping and regionalization is possible for hydrograph parameters—principally the height, width and half-width of the simple triangle to which the unit hydrograph is normally reduced. These dimensions represent the *peak-discharge rate* (Q_p) for the unit of rain (e.g. per 25 mm), the duration of the flood (time-base or Tb) and, critically, the time-to-peak (T_p) which is often used as the basic parameter of the hydrograph for relating to catchment characteristics. For instance, the *Flood Studies Report* (NERC, 1975), the British Isles 'bible', recommends:

$$T_p = 46.6 \, MSL^{0.14} \, S1085^{-0.38} \, (1 + URBAN)^{-1.99} \, RSMD^{-0.4}$$

where

MSL   = main stream length (km)

S1085   = slope (m per km^{-1}) between 10 and 85 per cent of main channel length.

URBAN = fraction of catchment with urban land use

RSMD   = a five-year recurrence interval daily rainfall with mean soil-moisture deficit subtracted,

and from it:

$$Q_p = 220/T_p$$
$$Tb = 2.52 \, T_p.$$

Wilson (1983) also describes how to synthesize hydrographs from these basic predictions.

4.4.6 Flood prediction, forecasting, and protection

In our introduction to this topic (Section 4.1) we discussed the danger of an excessively anthropocentric approach to extremes and perhaps, in philosophical mood, the advantages to be gained by more organic responses. On purely practical grounds hydrological management over much of the world is now attempting to grasp approaches in which the sensitivities of the broader basic environment to extremes are balanced before far-reaching and long-lasting anthropocentric solutions are implemented. However, prior to this perception of the need for balance, heightened by many unsympathetic schemes, the anthropocentric approach to floods had been to control them and to 'train' river channels to cope with them.

Thus a polarized case for the traditional approach would be to:

• rid agriculture of problems created by wet soils by 'getting the water away' from the land;

• rid settlements in attractive locations (such as fertile floodplains) of the hazard to their lives and work by containing floods in the channel.

We return to the more modern viewpoint in Section 4.4.7 but it must be admitted that in the case of one area of flood control; *dam-proofing*, there are more commendable needs for perfection. It is an obvious corollary of the natural variability in water resources (both in space and time) that large volumes of water storage are required. We have already considered natural storages and these should be optimized but there will always be a need for artificial reservoirs secured by *high dams*. Dams will always represent a hazardous venture in design and construction and, other than the seismic threat to their stability (best avoided by careful location), the risk of dam failure is best controlled by a detailed approach to flood prediction. If a dam can be adequately protected against being overwhelmed by flood flows (leading to erosion of earth dams or collapse of masonry or concrete) by a well-designed *spillway* (see Plate 4.2), society can often combine the benefits of water storage and downstream flood protection (Section 4.5.4).

There have been many dam failures since the first recorded one in Egypt in 2800 BC (Smith, 1972)—see Table 4.6. They continue to occur, often because suitably cautious legislation is not in place to enforce good design and regular inspection. With secure foundations, strong construction, and an adequate spillway—to discharge excess floodwaters away from the dam structure—it is possible to envisage dams lasting hundreds of years. The chosen design probability should reflect this longer view, especially where there are settlements in the danger zone below the dam with the risk of a high death toll if the structure collapses. The great utility of the unit hydrograph approach to spillway design on such dams is that a very rare rainfall, combined with the unit hydrograph for the reservoir catchment will allow prediction of, for example, the *Probable Maximum Flood*. The Probable Maximum Precipitation for such a calculation will come from meteorologists calculating the effects of the virtual 'collapse' of the moisture in the atmosphere above the catchment.

Plate 4.2. A full-width spillway in operation at the Clywedog Dam, mid-Wales, UK.

Plate 4.3. The depiction of rainfall patterns by radar for flood forecasters.

The hydrograph approach also offers those living or working in flood-prone situations the option of *flood warning* as an alternative to, or addition to, structural protection. For example, a radar rainfall scan (see Plate 4.3) can be combined with a computer model of the flood hydrograph to give real-time forecasts as a flood develops (Collinge and Kirby, 1987). Flood-warning systems require the involvement of the police, emergency services, and the local population because they need fail-safe, firm control, and preparation/practice for the vulnerable population. Widespread use of flood warning is unlikely except in highly sensitive locations and as a service to, for instance, livestock farmers in developed countries. It is more likely that, subject always to improved *cost-benefit analysis*, and with proper *environmental assessment*, structural protection will continue to be popular.

Table 4.6. Notable flood disasters resulting from dam failures

Date	Location	Fatalities and damage costs
11.3.1864	Dale Dyke, Sheffield, UK	250 deaths, 798 homes destroyed, 4000 homes flooded.
–.5.1889	South Fork Dam, Johnstown, Pennsylvania, USA	2000 deaths
2.11.25	Eigiau/Coedty Dams, Wales, UK	16 deaths
12.3.28	St Francis Dam, California, USA	420 deaths
2.12.59	Malpasset Dam, Frejus, France	421 deaths
9.10.63	Vaiont Reservoir, Italy	2600 deaths
5.6.76	Teton Dam, Idaho, USA	14 deaths, £400 million damages
15.7.82	Lawn Lake, Colorado, USA	33 deaths, £31 million damages

Sources: American Society of Civil Engineers (1975, 1988); C. Clark, (1983); N. Smith, (1972).

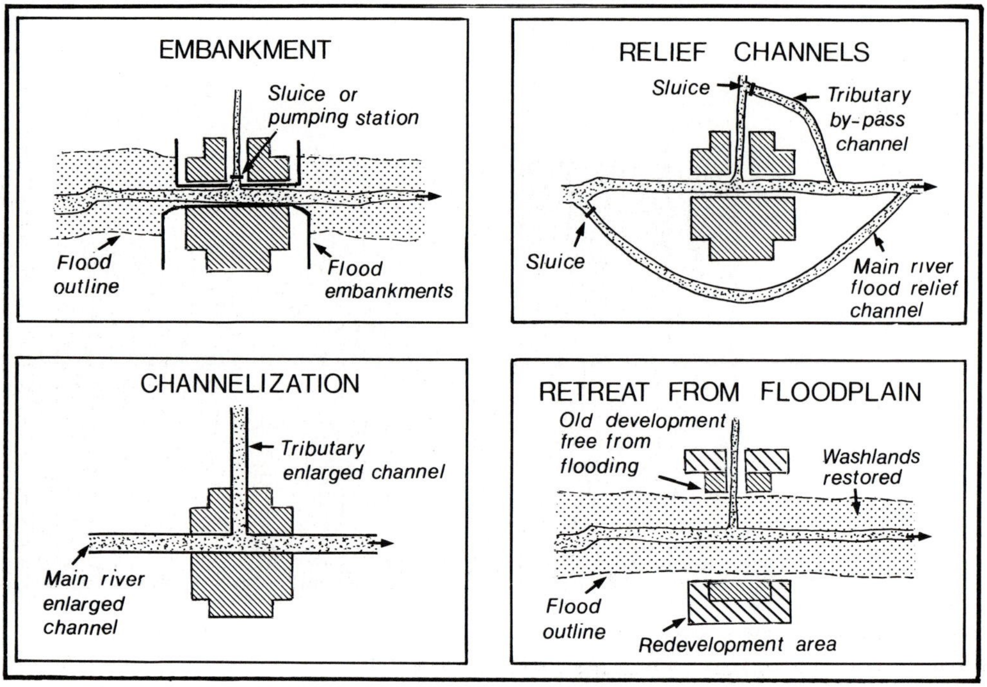

Fig. 4.11. Engineering/planning options in urban flood protection (after Nixon, 1966).

Fig. 4.11 presents cartoons of the major options for structural urban flood protection, including the option of the city withdrawing from the hazard in favour of washland/wetland restoration. Plate 4.4 illustrates the engineering works designed to spread water on to the floodplain in a controlled way in order to protect the city of Lincoln (UK). Such a decision has obvious benefits for wildlife in the river and its corridor may have amenity benefits, and also may clean up pollutants from the urban area (see Chapter 8).

4.4.7 Floods and nature: a conservation view

We have pointed to the resilience of biota in the face of what settled human societies refer to as damage; there is, however, another environmental implication of our demands for protection which is that the corollary of such protection is the loss of seasonal saturation or inundation of plant/soil systems known as *wetlands*.

We return to the proper scientific description and hydrological understanding of wetlands (which include areas of temporary open water) in Part II of this book but it is appropriate in concluding a section on floods to incorporate a simple statement of the interrelationship between flood control (by engineering means) and the regime of river extremes which would prevail in nature (Table 4.7).

Not only do structural schemes of land drainage and flood protection lower water-tables, reduce inundation, and disturb bed and bank habitats during construction but they must be regularly maintained. They are frequently accompanied by concrete or stone erosion-protection and they tend to collect sedimentation requiring removal (Plate 4.5).

Studies which have been carried out on structural flood protection schemes indicate a considerable impact on wildlife; fisheries may recover by natural restocking from outside the affected reaches but riparian and corridor species, especially birds, are affected by a profound alteration in their habitat (see Lewis and Williams, 1984).

Fortunately, engineers and conservationists have established a dialogue in most developed countries with the result that more harmonious and *ecologically acceptable channel designs* are now being attempted. It is essential to retain more diversity of riparian/corridor habitat, even if this is restricted to one bank or is in the interests of recreation rather than conservation (Plate 4.6); there are very considerable benefits for human amenity (see Gardiner, 1991).

Plate 4.4. Use of controlled washland flooding through sluices (foreground) to protect downstream urban area (Lincoln, UK).

4.5 Drought and drought hazard

In the world's *drylands*, the arid and semi-arid zones on Earth, drought may be considered normal in terms of ecosystem adjustments. However, in many other areas hydrological prediction is required to protect against *episodic* droughts; there is no obvious signal on the map of world climate as to where these will occur. There are geographical links within the drought problem: zones such as the wet highlands of the world, may be intimately concerned with drought management in adjacent semi-arid areas by supplying irrigation water.

As revealed in Chapter 1, life without water is impossible; thus whilst flood protection can run counter to the success of natural habitats (when conducted from a short-term human standpoint), drought protection is essential for most forms of life (though some have considerable but not permanent resilience), and the broad philosophical issue here is how close to the drought hazard should settled human societies develop. This is particularly relevant to the sophistication of the development (e.g. High Plains cities of the USA) and the density of population (e.g. resettlement schemes in Ethiopia); high levels of sophistication and high densities of population raise the stakes in the provision of water supply. High levels of hazard are compounded by high levels of risk. Environmental impacts may then be felt, not locally as in the case of flooding and

Table 4.7. Flood control and natural flow extremes: a conservation perspective

Aspect of flooding	Engineering control	'Natural' options; conservation benefit
Seasonal, waterlogging of soils and surface, standing water.	Cultivation and drainage systems at field level, requiring efficient arterial systems of removal for water surplus.	Retention of wetlands as habitat in key locations for temporary water storage. Exploitation of commercial benefits of wetland crops or recreation. (Water-quality benefits: see Chapter 7).
Inundation of agricultural floodplains.	Floodbanks and river draining give security of grazing or cultivation. Flows to coast accelerated, any resulting erosion controlled structurally.	Use of floodplains as temporary flood storage to benefit downstream sites. Operation of warning systems. Use of key wetland ecosystems as flood storage by pumping. Retain natural sinuosity and vegetation of channels.
Urban sites towards less sensitive rivers, communication.	Comprehensive schemes of river deepening, widening, and embanking.	Re-plan land use adjacent to classes such as recreation, parking. Warning systems. Off-river storage for supply.

Plate 4.5. Heavy structural channel grading for flood protection. Note lack of conservation/recreation features of banks and river corridor.

floodplains, but far distant in the supply zone as more valleys are dammed. Groundwater systems, of immense value during drought, can be mined for water without consideration for the rate of recharge. Some such aquifer supplies were in fact recharged in a previous climatic era and are unsustainable in terms of modern demands. As revealed by Beaumont (1989), a large number of world cities are in drylands and their populations are therefore exposed to considerable risk (see Table 4.8).

4.5.1 Time-scales and spatial scales of droughts

Flood episodes are short and can be created by intense local precipitation. In many climates there is a flood season in which unprotected land is flooded for weeks but within this period the individual episodes may be timed in hours (upstream) and days (downstream).

By contrast, droughts slowly build up. Whilst floods are caused simply by surplus precipitation,

Plate 4.6. Retention of conservation/recreation features (right bank) in urban flood protection: Shrewsbury, UK.

Table 4.8. Major urban dryland centres with more than 2 million people

Continent	Estimate date	City population	Urban agglomeration
Africa			
Cairo (Egypt)	1974	5 715 000	
Alexandria (Egypt)	1974	2 259 000	
Asia			
Peking (China)	1970	7 570 000	
Tientsin (China)	1970	4 280 000	
Teheran (Iran)	1973	4 002 000	
Delhi (India)	1971	3 287 900	3 647 000
Karachi (Pakistan)	1972	3 498 634	
Lahore (Pakistan)	1972	2 165 372	
America			
Los Angeles (USA)	1975	2 727 399	7 032 075
Santiago (Chile)	1970	3 273 600	3 350 680
Lima (Peru)	1972	2 833 609	3 303 523

Source: Cooke *et al.* (1982).

high melt-rates of snow, or by unusually rapid runoff (frozen ground), droughts are simply the result of the variability of precipitation, though different parts of the runoff cycle are affected at various times after the original shortfall in precipitation.

In Chapter 3 we defined a number of natural reservoirs within the ground phase of the hydrological cycle and drought definitions apply to each of these (Table 4.9). Thus, drought is neither simply a shortage of precipitation, nor is it of equal impact on the farmer, the fishery, or the reservoir operator. Finally the widespread occurrence of famine in the world, publicized in recent years as drought-related, is often the result of warfare, pestilence, and poor food distribution *in addition* to drought (e.g. Jansson *et al.*, 1987).

Certainly the phenomenon identified as 'desertification' in the 1970s is now acknowledged to have a large and complex web of causation beyond the increased rainfall variability accompanying climate change.

Clarke (1991), for example, defines four different causes of water scarcity:

- Aridity—a dry climate
- Drought—irregular episodes of dryness
- Desiccation—land degradation
- Water stress—overburdening of the supply.

One of the disappointing aspects of drought management to hydrologists is that whilst large areas (and hence high impact) and long time-scales (and hence the ability to provide warning) characterize

Table 4.9. Drought in the hydrological cycle: a reservoir approach

Hydrological variable	Characteristic scale of impact	Management response
Precipitation	*Two weeks* without rain often quoted as drought. In some areas a *winter* without snow has major impact.	Primitive ceremonials, Developed world: cloud-seeding (little used).
Shallow/surface soils and vegetation	At early growth stages drought impact quantitatively described (e.g. soil moisture deficit—see Section 3.4.5). May arise in a *few days*. Rates of evaporation (i.e. heat and wind speed also need to be considered).	Irrigation systems where crop valuable and assessment of frequency justifies. Elsewhere replant or lose crop.
Soil column and mature crops	Drought *season* consideration, plus health of crop at establishment, i.e. climate in previous season.	Choice of suitable crop/day farming techniques of equal validity to irrigation.
Groundwater	Many systems respond over a series of *years*; rate of abstraction increases memory of system.	Artificial recharge schemes; management of abstraction.
Rivers/surface reservoirs	Often dependent on *one season* of recharge but up to *three*, depending on supply source for river and size of reservoir.	Siting and development of reservoirs to balance human and ecosystem requirements/pollution control.

droughts, by comparison with floods little technical progress has been made towards protection against 'rare' thoughts since no mere structural devices are available and the options are mainly socio-political.

4.5.2 Defining droughts

The time-scales defining episodic drought hazards are those of the filling and emptying of reservoirs, both natural and artificial (Table 4.9). The shallow interception storage on plant canopies is dry for much of the time between precipitation events; the *upper soil* holds more water and therefore dries out more slowly with plant roots extending deeper in the soil column as it dries. By continuing to extend the reservoir analogy we pass through the very large *groundwater* store (which may keep streams flowing for months in drought) and into the realms of constructed reservoirs; these are also drought sensitive according to their capacity and the relationship between that capacity and the catchment area. Whilst small reservoirs may be depleted by use in a single dry summer (temperate conditions), the large modern reservoir may only be sensitive to two or three years of dry conditions. Within that period it is the successful rationing of the available supplies which can make drought protection a success. Once again, socio-political decisions are as important as information on the likely duration of the drought.

The sequence of impacts during a lengthy drought period is likely to run in the following order as reservoirs, both natural and artificial, become empty:

- Declining agricultural yields
- Loss of sensitive crops unless irrigated
- Widespread crop failure
- Death of livestock; failure of irrigation
- Death of forest cover
- General environmental and farming degradation.

From the statistical point of view a drought which has a long recurrence interval in its impact on crop yields (i.e. mainly affects the surface soil through precipitation failure) may be much less rare in its impact on groundwater levels (see Section 4.5.6); it is therefore very important to quote the probabilities of a particular drought in terms of which component of the runoff cycle is under consideration and the chosen duration for that treatment (e.g. is it lowest daily river flow or lowest 6-month flow volume which has been analysed?)

4.5.3 Droughts and river flows: biotic effects

It is generally considered, from an anthropocentric viewpoint, that low flows in surface channels are of most seriousness because they jeopardize the recharge of reservoir volumes or fall below river off-take schemes for human supply or irrigation. Even from such a viewpoint this is short-sighted; the pollution risk during droughts is increased through lack of diluting water from catchment runoff. Navigation and amenity uses of the river also suffer; even in the seasonally dry climate of the Mediterranean region the river amenity is poorly developed because of lack of interest in a stagnant or dried-up channel.

The *biotic effects* of drought are even more important and are correlated with increased concentrations of pollutants, competition for food, and increased predation. Water temperature also tends to rise in droughts because flow is less turbulent and river water becomes less aerated; demands on the *dissolved oxygen* content are increased by the density of the biota in the remaining flow and by the increased rate of biochemical reactions. Finally, the area of substrate available for shelter or breeding activity shrinks in proportion to the reduction of river depth and width. This reduction may also have severe effects on agriculture as cattle can no longer reach water but can cross dried-up or shallow channels which, under normal conditions, form 'wet fences' at property boundaries.

An increasing research effort is being made by freshwater biologists to calibrate the river flow and quality needs of the river-channel habitat so that the impacts of both natural extremes and river regulation can be predicted (see also Section 5.8.4).

For these reasons, the management of drought flows is of increasing hydrological and political interest as the competing needs of water storage, direct supply, and river biota are considered. The use of river regulation as a supply strategy has advantages in these circumstances since, on its way to the point of human demand, the released water can serve a conservation, fisheries, and amenity function. Groundwater-to-river transfers are also a useful strategy in such circumstances.

4.5.4 Reservoirs, droughts, irrigation, and downstream effects

Human beings have been damming streams for nearly 5000 years (Smith, 1972). The earliest-known dam was apparently destroyed by poor flood protection not by environmental protests! As we have just observed, drought protection must be applied, not only to the direct beneficiaries of reservoir construction (human consumers, industry, power, agriculture) but to the river landscape, habitat, and ecosystem downstream. One of the earliest legal principles of our use of the common water resource was that of *riparian rights* which states that:

[A proprietor of land adjoining a river] has the right to have it come to him in its natural state, in flow, quantity and quality and to go from him without obstruction. (Wisdom, 1979: 83)

It was in the earliest days of manipulating river flows in England that this principle became translated into working practice. As Gustard *et al.* (1987) reveal for the United Kingdom, deliberate storage of river flows to supply power to mills or drinking water to a rapidly urbanizing people, led to disputes with downstream users. The terms *compensation water*, and latterly *guaranteed flows*, became written into English law to ensure both a fair distribution of the common resource and environmental protection for the river channel downstream of the weir, dam, or other offtake. Even where the use of water is non-consumptive (i.e. where a return of an equal or little-reduced volume is possible as in hydropower and from low-intensity human consumption) there are principles of flow regime to consider. Increasingly, too, there are important issues of water quality below reservoirs, with releases representing a power to dilute polluted river flows. There is potentially a double impact for reservoirs which supply urban consumers in the same catchments; those consumers return polluted sewer flows to the river which, however, lacks the dilution of rural, upstream flows because these are stored in the reservoir! A very large proportion of summer low flows in some UK rivers has been contributed from sewer systems.

Similar arguments apply to irrigation; under the same conditions where headwater (largely unpolluted) flows are stored for use in irrigated agriculture, both the quantity and quality of the *return flows* to the river are diminished because transpiration concentrates the remaining soil solution. In *salinized soils* the concentration of salts is so high in the upper soil layers that productive cropping is no longer possible and the irrigation scheme is ruined. Drainage water leaving irrigated areas also contains elevated concentrations of dissolved salts; in the Colorado River of the Western USA salt concentrations doubled in the 1970s making the river's use

for agriculture or human consumption in the lower basin impossible. International problems have arisen between the USA and Mexico, that nation being a downstream user of the river's water (Graf, 1985).

The environmental influence of river-flow volume and timing as a result of reservoir storage and regulation is now receiving special attention in water-resource schemes (Petts, 1984).

We return to this problem in Chapter 5 but, for the moment we include the simple influence of reservoir-regulation flows on the downstream system. Use of the river channel system to carry water volumes to alleviate droughts is certain to have impacts on:

- The channel itself, including sediment transport
- Chemical content of the river, including pollutants
- Fisheries/biota
- Navigation
- Amenity
- Local climate.

4.5.5 Droughts and irrigation: an international view

A large proportion of mankind's exploitation of soils and of natural biotic activity occurs in the upper soil. Consequently, because of its drought sensitivity many regions of the world rely on agricultural irrigation, to yield crops and create the basis for settlements. A recent world survey of the existing and potential use of irrigation (Higgins *et al.*, 1988) revealed that around 15 per cent of world agriculture is irrigated (see Table 4.10) but that this land produces 40 per cent of world food. Clearly, properly planned and managed irrigation is capable of profoundly improving the lot of mankind on Earth; the paradox is that poor irrigation schemes are now also a major cause of environmental degradation through salinization of soils, break-up of settlements, and downstream impacts of both dams and irrigation return flows.

Simplistically, irrigation is the reverse of drainage; indeed the same pipe networks can function in reverse if required. Modern 'trickle techniques' employ small perforated pipes laid between crop rows to take controlled amounts of water directly to where it is needed—the roots. Sprinkler irrigation is also highly controllable and efficient but entails interception losses. However, the technology of applying water to soils is generally much simpler in most heavily irrigated areas: the traditional flooding from canals is most widespread. This runs the risk of erosion unless fields are levelled, an expensive

Table 4.10. Continental distribution of irrigated area, 1984

	Irrigated area (10^6ha)	Percentage of world total
Asia	136.865	62.30
North America	20.461	9.31
Soviet Union	19.485	8.87
Europe	15.710	7.15
Africa	10.390	4.73
South America	7.979	3.63
Central America	6.914	3.15
Oceania	1.869	0.85
Developing countries	157.198	71.56
Industrial countries	62.475	28.44
WORLD	219.673	100.00

Source: Higgins *et al.* (1988).

process, and also encourages salinization unless drainage schemes are also constructed. Furrow, ditch, or canal irrigation is a modified form of flood irrigation (see Fig. 4.12).

Irrigation design, rather like the reverse of drainage, requires the hydrologist to know the likely crops to be grown and their sensitivity to drought. Optimum water use by crops varies throughout the crop cycle; thus, efficient water use reflects the need to restore calculated soil-moisture deficits at their critical magnitude. For example, in the UK the following principles are employed in horticulture:

- Brussels Sprouts: Give 25 mm after transplanting. Single application of 40 mm when deficit >40 mm (when sprouts 15–18 mm in diameter).
- Carrots: Give water before drilling: 25 mm in April or 50 mm May–July. When roots grown add same amount as soil moisture deficit when this reaches 25 mm or 50 mm (according to soil type).

Other examples are listed by the Ministry of Agriculture, Fisheries, and Food (1982). Of course such precision requires growers to calculate, measure, or receive advice about deficits and then to own equipment sensitive enough to apply the right amount. Such a rational approach is not widely available and would be impossible in much of the developing world; in the semi-arid zone it is fortunately only necessary to make seasonal estimates of crop requirements: rain is unlikely!

4.5.6 Predicting drought and climatic change

In introducing a treatment of drought we differentiated between humid climates affected by episodic droughts (or by the needs of neighbouring drylands) and the arid/semi-arid lands affected by virtually perennial drought.

In the former the hydrologist's approach to drought prediction takes two forms:

- The assessment of the *probability* (and hence return period) of the drought extreme.
- The analysis of *river low-flows* or the rates of fall of the water-table in groundwater systems.

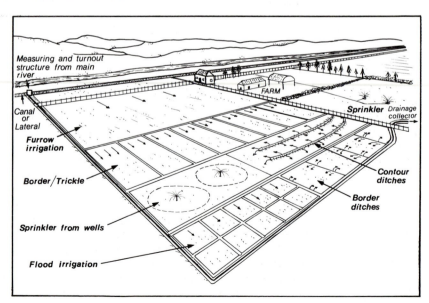

Fig. 4.12. Major irrigation techniques (unrealistically assembled on one farm!).

For probabilistic approaches, there are clearly a number of candidates for analyses, e.g. the minimum river flow during the drought, the period without rain or the drought average rain, snow-depths (for mountain-fed reservoirs), evaporation rates, and so on. Each of these elements can be treated by virtually the same statistical analyses as are flood-flow data; the results often reveal differential probabilities according to the hydrological component assessed and the duration chosen (see Table 4.11). Each can be accorded the tabulation and plotting routines outlined for floods data in Section 4.4.3, with exactly the same caveats concerning the choice of distribution and the relationship between recurrence interval and risk.

Table 4.11. Return period of the 1975–1976 UK drought for different river-flow stations and durations of flow

Record length (years)	Station	Duration in months				
		1	3	6	8	18
63	Yorkshire Bridge	25	25	10	7	15
44	Bedford	50	20	20	20	20
94	Teddington	90	50	30	35	20
56	Bewdley	200+	200+	35	7	30
69	Caban Coch	100	200+	70	8	40
50	Stocks Resvr.	10	100	9	4	9

Source: Modified from Hamlin and Wright (1978).

The statistical analysis of river low-flows also yields variable results according to the chosen data-base: there will be a regional variability based on the inherent patterns of rainfall and evaporation but one profoundly affected by both natural and artificial storages. Clearly a reservoir-regulated river will not be suitable for such analyses, or will need its flows to be corrected or *naturalized* by the sub-traction of artificial releases. Other elements of an artificial flow record are harder to detect, e.g. water pumped out from mines and into rivers and water passing in the opposite direction into irrigation.

In the case of natural storage it is possible to relate the analysis of low-flow patterns to soils or geology within the catchment concerned. First, careful analysis must be made of the baseflow characteristics of the catchment in question; the quickflow component, selected for unit hydrograph analysis, is now discarded and the *recession curve* of baseflow averaged over the annual flow record to produce a baseflow proportion of total flow. Section 3.3.3 shows how the baseflow index is derived and some sample UK values. From this basic parameter

(rather like T_p in the unit hydrograph) prediction equations allow calculation of a full range of low-flow statistics, namely for different durations of time and for different return periods of recurrence.

The recurrence interval of some definitive parameter of drought may be predicted, though this becomes more difficult under conditions of changing climate. Nevertheless, *rainfall variability* can be mapped at both global and national scales (Fig 4.13(a) and (b)). Where long rainfall records exist they can be analysed probabilistically to yield measures of the likelihood of drought. However, much of Africa appears to be 'locked into' drought as a possible consequence of climate change. The best possible aid for famine-torn nations affected by lengthy drought is some form of forecasting and warning response to the hazard.

Hope is growing amongst climatologists that there are exploitable connections (*teleconnections*) between, for example, sea temperatures, ocean currents, and the circulation of moisture in the atmosphere. Satellite observations may then be capable of triggering global forecasts of serious drought; the same observation system can incorporate monitoring of land-use change which can be a major intensifying factor through its effects on atmospheric processes (precipitation/evaporation) and on land degradation or neglect.

4.6 Runoff extremes outside the humid-temperate zone

We have already referred to the importance of the form, seasonality, and intensity of precipitation in controlling hydrological response (Sections 2.10 and 3.8). When dealing with extremes we also need, of course, to consider lack of precipitation—the major cause of droughts. We also need to consider the relative role of human agency in producing extremes; the development process is said to increase the vulnerability of human populations to both drought (e.g. within the process known as desertification) and floods (for instance where dams are damaged or burst).

Beginning with floods, Hayden (1988) provides a useful world-climate classification on the basis of flood-causing mechanisms; he considers two basic atmospheric conditions—barotropic and baroclinic—together with the dichotomy between rain and snow as the causal precipitation, and maps seventeen flood climates. Hirschboeck (1988) translates

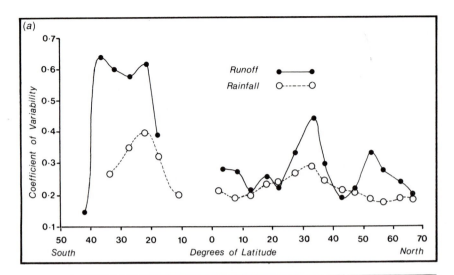

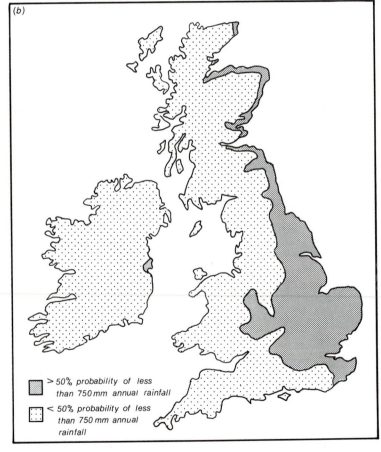

Fig. 4.13. Rainfall variability; (a) by latitude (including runoff) (after McMahon *et al.*, 1987); (b) within the British Isles.

this global treatment into one based upon scales more suitable to the study of floods in individual river basins. She is particularly keen to isolate those meteorological phenomena which either intensify a normal precipitation system or lead to the creation of atypical patterns, such as those prevailing during the 1972 Hurricane Agnes event in the eastern USA—said to be the region's greatest-ever natural disaster. During Agnes the entire North Atlantic circulation was anomalous, 'trapping' the hurricane over the eastern seaboard. Hirschboeck suggests that a *hydroclimatological approach* to flood frequency analysis should allow for the varied climatic origins of individual events (see also Fig. 4.9) by separating populations of flood peaks deriving from, say, convective, frontal, and orographic rainfall activity within the same basin. In a biblical reference she distinguishes basically between 'Joseph effects' (long wet or dry spells) and 'Noah effects' (short, sharp events).

The most comprehensive contemporary guide to floods under differing climatic conditions was compiled by geomorphologists in an attempt to trace the feedback between flood power and fluvial features (Baker *et al.*, 1988). Schick (1988) contributes considerable first-hand experience of gauging *floods in semi-arid areas*, some of whose features were introduced in Section 3.8. He refers particularly to the phenomenon of translation losses down the wadis and other ephemeral channels of the semi-arid environment. Of nineteen events studied over a 1 km reach of the Nahal Yael (Israel), nine increased in peak discharge downstream (the predictable result in 'normal' river basins), seven decreased, and two were equal. Maximum losses into the alluvial bed of the wadi were 8 l m^{-2}. The rainfall data for this same catchment indicate the importance of the extreme intensities in generating rapid-runoff responses. Over a 17-year period half the total rainfall of 450 mm fell at intensities of >14 mm hr^{-1}. and a third of this fell at intensities of >2 mm min^{-1}. It is not at all unusual for Saharan rainfall stations to record an average annual rainfall of 20–40 mm but to experience *daily* falls of 30–50 mm.

Sutcliffe *et al.* (1989) describe the flooding of Khartoum (Sudan) from the August 1988 storms over the Nile and Atbara floodplains; these delivered between 64 mm and 210 mm of rainfall (more than the annual average) in one day, with 50 mm falling in the first hour. These authors make use of satellite imagery to interpolate between the sparse rain-gauge network. Schick also refers to the critical role to be played in future by remote sensing of semi-arid zone flood events.

The Khartoum flood-wave, which made 1.5 million people homeless, moved down the Nile at an average velocity of 112 km d^{-1}, but many flood-prone semi-arid settlements do not have such protection of distance. Las Vegas, for example, is one of a number of US 'sun-belt' settlements built on an alluvial fan; the consequence of this site is frequent flash-flooding of many of the notable downtown properties.

Church (1988) develops the special features of *nival floods*. The largest flood recorded on Earth was possibly the outburst of glacial Lake Missoula into west Montana, USA between 16 000 and 12 000 years ago. Church quotes a peak discharge of 21 000 000 cumecs! The geomorphology of a whole region was sculpted by this one event. In many countries, however, it is the physical obstruction of channels by ice, rather than the failure of the ice which forces up the stage, particularly where, in cold-zone cities, bridges and other structures become blocked.

Box 4.2. *Classification of cold region floods*

1. Hydrometeorological origins
   (a) Snowmelt/glacier ice melt
   (b) Rain on snow
   (c) Rain after thaw
2. Channel obstruction
   (a) Winter or thaw ice blockages
   (b) Icing of channels from seeps and springs
   (c) Glacier outbursts ('Jokulhlaups')
3. Failure of glacial moraine dams or blockage by solifluction/ landslides

Source: Modified from Church (1988).

Even in the temperate zone snowmelt floods are important. For example, within the UK, Johnson (1975) lists a majority of Britain's largest catchments whose highest recorded flood was dominated or at least influenced by snowmelt.

Moving to another extreme, i.e. of the *humid tropics*, Gupta (1988) tabulates the almost unbelievably heavy daily or circa-weekly rains of the humid tropics (Box 4.3—see also Box 3.2). Such falls dominate the flood behaviour of catchments in parts of east and south-east Asia, humid India, Madagascar, the Caribbean, and northern Australia. They are a feature of monsoonal or tropical storm systems but are often intensified by orographic effects.

Box 4.3. *Some flood-producing rainfalls in the humid tropics*

Date	Location	Amount (mm)	Cause
16.3.1952*	Cilaos, Reunion	1870	—
11.9.1963	Pai Shih, Taiwan	1248	Typhoon
22/23.1.1960	Bowden Pen, Jamaica	2789	Frontal
9–16.6.1876	Cherrapunji, India	3388	Monsoon

*part of an 8-day total of 4230 mm
Source: Modified from Gupta (1988).

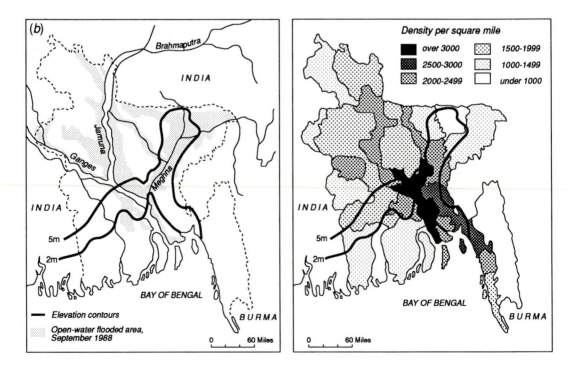

Fig. 4.14. Bangladesh's flood problems from (a) the Ganges–Brahmaputra catchment (after Brammer, 1990*a*); and (b) from sea-level rise and delta retreat in relation to river floods and population density (after Milliman *et al.*, 1989).

The Cherrapunji rainfall station lies in the huge Ganges-Brahmaputra basin which produces such ravaging floods in Bangladesh. Although Bangladesh comprises only 7.5 per cent of the basin its low-lying topography and dense rural population, heavily dependent upon floodplain cultivation and irrigation, render the hazard extreme (Fig. 4.14). Brammer (1990a) describes the flooding of 1987 and 1988 which resulted from different causal components since the river exhibits peaks due to snowmelt in the Himalayas, 'flash floods' from the foothills, and the monsoon season's deluges (which make Cherrapunji's record falls). At its record peak on 30 August 1988 the Brahmaputra flood was 50 km wide. Proposals to protect Bangladesh (described by Brammer, 1990b) focused at first on structural solutions—mainly embankments serving the interests of urban populations, for example Dhaka. However, alternatives are under investigation such as dam construction upstream (which might 'cream off' 10 per cent of the flood volume but be very susceptible to sedimentation problems) and groundwater abstraction to create storage in the soil for monsoon rains. With a forecast of global sea-level rise the nation is also at risk. Milliman et al. (1989) warn that Bangladesh faces the triple problems of sea-level rise, delta subsidence, and reduced delta growth (as dammed rivers reduce the sediment flux). Putting these together, by 2100 there could be a 3 m rise in relative sea-level, a 2 km retreat of the shore, 26 per cent reduction in habitable land with 27 per cent of the population displaced, and GDP reduced by two-thirds.

Completing the range of flood types in the 1988 review, Costa (1988) tabulates the available information about floods caused by *dam failures*, both artificial and 'natural' dams (i.e. those formed by landslides across valleys or ice blockages). The lax legislation on dam safety in the USA makes it a centre for the collection of dam-break flood-discharge information. Gathering such information requires great skill in reconstructing the peak discharge from geomorphological and other 'damage'. Information on the cost to society, including loss of lives, is more readily available. By comparing flood peaks with the information on the type, location, and shape of the dam breach it is possible to develop predictive equations for the catastrophic flood peak (Q_{max}) resulting from dam failure, the simplest of which is:

$$Q_{max} = 48 \ H^{1.63}$$ (where H is the height of dam in metres)

Turning to the drought extreme, whole books are written on aridity (e.g. Beaumont, 1989; and Agnew and Anderson, 1992); the topic is particularly suited to geographical breadth because the crisis of aridity is one induced by human behaviour. One third of the Earth suffers from episodic and punishing drought and whilst it houses only 9 per cent of the population the arid/semi-arid zone is one of rapid urbanization (36 millions live in Cairo, Karachi, Lima, Beijing, and Tehran), and of appalling rural poverty and famine. Irrigated agriculture is, of course, the answer to the livelihoods of all these people but hydrologically the technology of irrigation is still poor, with problems such as salinization wiping out as much irrigated food production as comes on stream each year in new schemes. The fact that irrigation is not the triumph it could be is indicated by the sensitivity of food production to drought in a range of countries: Mexico 12 per cent, Australia 31 per cent, and the USA 20–30 per cent—apparently a case of 'the bigger they come, the harder they fall'.

5. Artificial influences on runoff volume and timing

5.1 History of concepts and experimentation

Chapter 1 briefly examines the history of hydrology as a science whilst Section 2.3 adds the use of catchment experiments as a device for accruing data on components of the hydrological cycle.

For the purposes of this chapter we need to particularize the development of a perception by hydrologists that artificial influences wrought by the *development* of both water and land resources by human society, might control the processes and patterns of rainfall, runoff and storage in the hydrological cycle.

Such a perception of downstream influences clearly has a long antecedence in the case of direct manipulation of surface waters: witness the riparian rights and by-laws stipulating good practice which date in the UK to the medieval period. It has been much more difficult to bring into both scientific and popular purview the relationship between *land use* and surface water (both quantity and quality, and extending to the effects of land *management*). There exists a further challenge, namely to appreciate the relationship between surface land-use activities and the quantity and quality of *groundwater*.

Whilst Table 2.3 lists some of the important catchment studies which have developed our concepts of the hydrological cycle, and also *revealed the influence of land use* (notably forest cover) on runoff, it does not chart the progress of policies on the *management* of this interrelationship; this function is served by Table 5.1.

We can identify an early phase in which environmental extremes such as *floods* and *droughts*, together with river instability, led to state policies on afforestation. Later attention turned to *water resources* as well, and in the last two decades to water *pollution* from both point sources and diffuse sources (see Chapter 8). Finally, considerations of freshwater *habitats* in downstream locations have also threaded the cause–effect chain to upstream land use and management (see Newson, 1992).

Table 5.1. The relationship between land use and runoff quantity/quality: progress of policies

Date	Agent/Key players	Outcome
1215	Louis VI France	'The decree of water and forests'
1342	Swiss provincial government	Community forest as protection against avalanches
1841	Surell's book, *Etude sur les torrents des Hautes Alpes*	French gave Forest Department responsibility for controlling Alpine torrents
1864	George Perkins Marsh, *Man and Nature* (240 pages on *The Woods*)	Shaped public opinion on development and water resources/extremes
1930s	United States Soil Conservation Service and Flood Control Acts	Control of small catchments as part of agricultural recovery and downstream flood reductions
1960s	Research on climate/ runoff of cities; process studies on urbanizing areas	Planning and water authorities increasingly liaising on proposed developments
1970s	Research on forest and farm drainage and floods	Codes of land management practice
1980s	Research on groundwater pollution, diffuse pollutants, and effects on habitats	Legislative zoning of land use and management controls

The translation of research into practice is normally:

- Gradual, punctuated by technical support through hydrological design.
- Incomplete—there are compromises to allow development of some form to proceed.
- Voluntary in the main, with an emphasis on good practice in individual developments.

5.2 A return to processes: points and extent of intervention

As already described, the hydrological effects of direct intervention in surface water systems (e.g. of damming a river to create storage) are easy to imagine, though research has revealed many complexities and sensitivities which have influenced our perception of the spatial and temporal extent of impacts: these are covered in Section 5.8.

We therefore give greater highlight here to the often hidden processes of runoff generation considered for natural slopes and channels in Chapter 3. A useful way to catalogue the potential for human impact is to superimpose our knowledge of land-use and management effects on a basic-process diagram from Chapter 3 (see Fig. 5.1). This illustration can-

not, however, do full justice to the complete range of influences, some of which occur indirectly (e.g. on precipitation). Table 5.2 fills out the range of potential impacts and the reader is referred to the many compilation volumes of research results from individual land-use and land-management studies (e.g. Hollis, 1979; Solbe, 1986; Arnell, 1989).

5.3 Fundamental effects: volume and timing

The influence of mankind's use of land and exploitation of other resources on runoff is still debated both between research hydrologists and, where technical consensus is reached, between scientists and politicians. Where impacts are upon

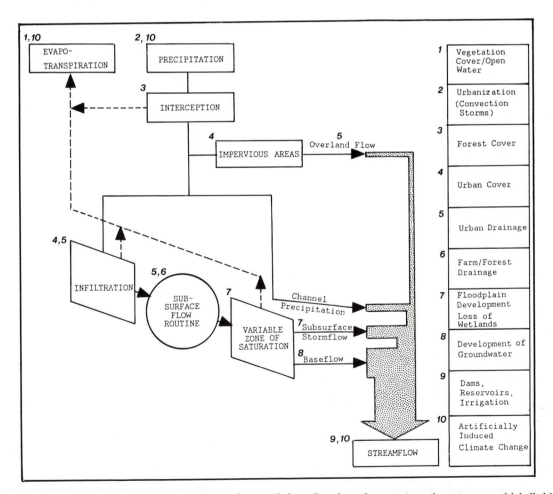

Fig. 5.1. A simple 'tank' model of the runoff cascade extended to allow for a dynamic contributing area and labelled by relationship of components to artificial modifications (land-use and water development).

Table 5.2. A partial catalogue of potential effects of land use and land management upon the catchment hydrological cycle

Component of cycle	Land use/management technique	Impact and reference
Precipitation (esp. convection rainfall)	Urbanization; urban heat island/pollution	Increased thunder-storm rainfall within urban areas (Atkinson, 1979)
Snow accumulation and melt	Forest cover pattern — clearings and felled areas	Clearings accumulated snow; snow sublimates from canopy but melts more slowly beneath than in open (Swanson, 1973)
Interception loss	Height of crop, exposure of canopy (also applies to spray irrigation of vegetables)	Afforestation reduces annual runoff in wet/windy climates (Newson and Calder, 1989)
Surface detention and Hortonian surface flows	Cultivation, drainage, change of soil and surface textures, and vegetation roughness	Urban surfaces have little detention capacity (Hollis and Ovenden, 1988); ploughing effects (Reid, 1979)
Subsurface drainage and groundwater recharge	Drainage/cultivation. Urban drainage and storm sewers	On naturally-wet soils drainage increases runoff rate in storms (Robinson, 1986). Reverse effect in more permeable and cracked soils (Section 5.5). Urban systems intensify surface effects of urban cover
Riparian zone and channel.	Flood protection/intensive landuse	Reduced storage capacity and channel form roughness (Brookes, 1988)

water resources or hazards to communities (e.g. pollution from industrial development or accelerated flood runoff from pavements and sewers) remedial options, on cause or effect, are taken up once proof is established. However, the scope of environmental hydrology is such that public concern is now equally strong (if largely lacking in economic 'clout') for organisms within the *freshwater ecosystem* or for the *aesthetics* of river systems. Since these concerns introduce new sensitivities and subtleties this section is designed to separate, in the case of runoff patterns, the effects on timing of runoff (Table 5.3). A further aspect of timing is that of delay between land-use change, its hydrological impact, and any recovery mechanisms which might mitigate impacts in the longer term.

To consider, first, volume and timing as artificially separate aspects of runoff, consider a plot of saturated soil during a rain storm. Much more runoff will occur in terms of *volume* than when that soil is dry, permitting a proportion of the rainfall to infiltrate. If we now dig an open, smooth-walled ditch down the plot and repeat the storm under the same conditions, runoff volume remains identical but it leaves the plot more rapidly using the facility

Table 5.3. Partial separation of the impacts of land use on volume and timing of runoff

	Volume	Timing
Vegetation change (without management) e.g. fire, traditional felling	HS	GD
Crop change from natural vegetation requiring cultivation/drainage	OS	OS
Urbanization—change to impermeable surfaces	HS	GD
Urban drainage and sewer systems	OS	HS
River basin development and channel regulation	GD	OS

Key:
HS = Highly significant; OS = Often significant; GD = Generally detectable.

of the hydraulically smooth drain rather than the rougher soil surface (especially if vegetated).

The situation is shown in Fig. 5.2 which also indicates the resulting 'flood' hydrographs in cartoon form. Our problem in interpreting empirical findings, even from the most heavily controlled experiments in the field, is that runoff volume and

Fig. 5.2. Influences of forest/agricultural land-use and management on the volume and timing of flood-event runoff.

Impacts are illustrated (below) by cartoon streamflow hydrographs for flood events.

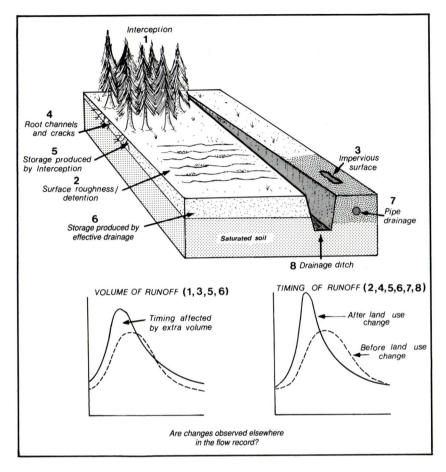

timing are seldom uniquely affected by a land-use development or management strategy. The case studies presented below make obvious the need, therefore, for extreme care in disentangling the precise route of cause and effect. To over-simplify once more, the increased velocity of runoff from our plot may reduce depression storage or surface-evaporation rates or may be part of a crop change; each of these impact also on volume. Arguing another way, if the saturation of the soil leads to a greater depth of surface runoff from the plot the effective roughness or friction to flow will be reduced, increasing the velocity of runoff and shortening the time base of the resulting 'flood'.

It is one of hydrology's greatest experimental problems that such complicated cause-and-effect chains are best elucidated by research in small catchments, whereas impacts are of greater relevance in much larger river basins. Downstream from the source of the effects the following phenomena blur detection:

- *Natural smoothing* as the upstream flow change is routed through storage in the channel,
- Interplay with *other artificial effects* which may have an equal but opposite effect (e.g. downstream abstraction versus upstream drainage). Acreman (1985) detects such interplay between afforestation in two parts of a single basin.

To these may be added the complication that many land-use and land-management changes are temporary or cyclical. Where an impact is detectable, even spectacular, it may be short-lived before previous conditions, or an altered but stable condition, are regained. Such a behaviour is exemplified by the sediment yield in UK rivers draining semi-natural moorland which have been afforested and then felled after the first crop cycle (see Fig. 5.3).

A theoretical framework for such changes through time is provided by the *threshold behaviour* of many natural systems (see Section 4.3). We have considered rapid, temporal changes as part of the

Fig. 5.3. The influence of coniferous plantations in the British uplands on sediment yield throughout the crop rotation (*c.* 50–70 years) (G. J. L. Leeks, personal communication).

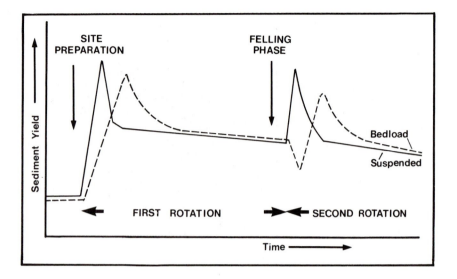

behaviour of natural extremes but similarities between Fig. 4.3 and Fig. 5.3 will indicate that artificial hiatuses may be modelled in a similar way. Inevitably there is a 'collision' of the natural and artificial threshold explanations of, for example, the widespread erosion and flood problems encountered throughout the historical period in mountainous areas (e.g. currently in Nepal and the Himalayas, but formerly in France and the Alps). Separating the two possible causes, of natural extremes and artificial destabilization, is also essential in improving *development techniques* in the third world and for mitigating the effects of *climate change*.

Within relatively stable climate regimes a separation of threshold events and their causes is often important as part of *environmental assessment* procedures. It can be argued by developers that, whilst their proposals for land-use changes are profound and the early impacts predictably large, society will eventually benefit as those impacts become reduced. The critical judgement is, of course, whether there is to be a threshold reaction and how predictable is the recovery towards existing conditions. There is an increasing awareness that natural and artificial threshold effects can overlap, as in the case of desertification, where drought is intensified by changing land- and water-resource developments.

For the remainder of this chapter we will assume that land-use change is the main external stimulus to hydrological change and that hydrological change may be gradual, or progressive, rather than rapid and threshold. Unfortunately, even the longest-running hydrological land-use experiments

have not yet yielded much information on these longer-term implications.

5.4 Forests and water

It was the perception that loss of natural forest cover had a profound and deleterious effect on rivers which paved the way for an expansion of scientific hydrology. Early environmentalists, including some who influenced public policy, had strong views on the need to retain forest covers to modulate flows, moderate extremes, and reduce erosion (Table 5.1). For example, Aldo Leopold in the USA wrote of the pure clear Gavilan River and its biotically stable natural forest catchment:

It was here that I first clearly realized that land is an organism, that all my life I had seen only sick land, whereas here was a biota still in perfect aboriginal health. (Flader, 1987: 16)

Leopold was to have a profound influence on the land management practised by the United States Forest Service. However, the impact of forestry on the hydrological cycle is not restricted to *natural forest covers* and their loss during settlement/development. Whilst the current onslaught on tropical forests will have a profound influence on tropical atmospheric moisture, profiles, and on river quality (Shuttleworth, 1991; Ross, Thornes, and Northcliffe, 1990) it is to the replacement *forest plantations* of the long-settled humid (especially mountain) zones and their management cycle that research hydrologists have directed attention in the twentieth century. A pioneering example in the UK was provided by the late Frank Law who established a forest lysimeter at Stocks Reservoir in the 1950s (Law, 1956)—see Plate 5.1.

Plate 5.1. The forest lysimeter operated in the 1950s by Frank Law at Stocks Reservoir, UK.

5.4.1 Elements of forest land use

Natural forests are generally long-lived and represent the climax vegetation for much of the humid zone where elements of the river environment are important to humans for a variety of reasons and to similarly long-lived freshwater ecosystems. Plantation forests are often, by contrast, the first managed crop for a long period on land hitherto labelled as surplus to development. In the case of upland forests in the UK there has been a 4000–5000-year gap in forest land covers and for the first time since prehistoric settlers felled the slopes, the 'moorland' soils are drained, ploughed, planted and roaded on a *cropping cycle* of just over half a century in duration.

Maitland *et al.* (1990) have suggested stages of UK plantation forestry, in the case of upland conifer forests, in terms of environmental impacts on freshwater ecosystems. A summary is provided by Table 5.4. The table indicates that it is not possible to speak of 'forestry effects', even within the narrow confines of plantations in the uplands of one nation. Instead, there is a complicated set of impacts arising partly from the use of the land for a tall crop, and partly from the cycle of management strategies used to cope with the relatively hostile environment in which the trees are now grown. Whilst trees were the climax vegetation of these hills 4000–5000 years ago, the deterioration of soil fertility since their loss and the change of microclimate signals the dangers of exploiting natural forests out of existence as pointedly as the contemporary loss of tropical forests.

The activities shown in Table 5.4 can clearly be adapted to minimize environmental impacts; to this end codes of practice have been published in several parts of the world (see Bruenig, 1986; Ponce, 1986; Forestry Commission, 1988). These codes encourage

Table 5.4. Forest land use: the plantation cycle and effects on freshwater ecosystems in the UK uplands

Stage	Characteristic management	Impacts
Ground preparation	Ground preparation includes clear-felling previous rotation but mainly ploughing, draining, and weed control. Road network revised or constructed.	Exposure of bare soils, change of albedo, dewatering of storm runoff, early erosion.
Planting to canopy closure	Pesticide and fertilizer applications, further road developments.	Danger of pollution incidents and eutrophication of standing waters.
Mature crop	Little management until thinning for pulp-wood at 25–40 years.	Interception loss of precipitation and collection of dry pollutants on canopy. Ditch erosion may continue.
Harvesting	Felling of forest compartments according to planting date and timber yield. Removal of selected timber, residual material (branches) left behind.	Dangers of soil compaction, channel damage, road damage impacting on streams. Waste timber materials rot and may destabilize slopes and add nutrient chemicals.

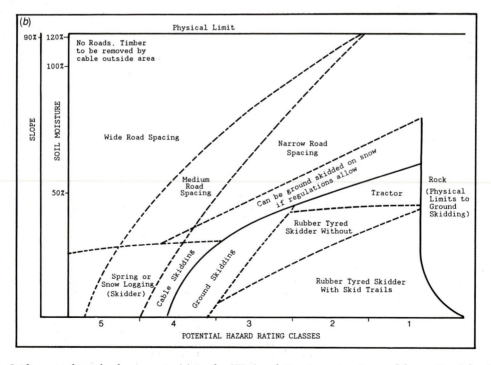

Fig. 5.4. Guidance to forest land managers: (a) in the UK, in relation to preparation and harvesting (after Forestry Commission, 1988) (letters key to specific practical guidance); (b) in British Columbia, in relation to logging techniques.

Table 5.5. Regional process controls in forest hydrology

Regional element	Effects on processes
CLIMATE	
Rainfall regime	Low intensities/long durations favour interception if other conditions are right.
Temperature/sunshine	Higher radiation inputs stimulate physiological water use via transpiration. Radiation climates essentially different from advection climates.
Wind speed	Advective energy and ventilation favour interception but also high rates of physiological stress and hence high transpiration where trees deeply rooted in wet soils.
TREES	
Species	Often irrelevant but fast growers use more water if available.
Stomatal control	Often superior to genetically produced agricultural crops.
Rooting depth	Critical control on transpiration rate in dry conditions, especially if soil has good moisture storage (e.g. phreatophyte).
Spatial scale of forest	Controls edge effects v. core effects.
SOIL/GEOLOGY	
Permeability	Controls water use and need/effectiveness of cultivation and drainage.
Chemistry	Controls impact of acid deposition and need to apply nutrients/other chemicals.
MANAGEMENT	Virtually complete control except over crop height and climate.

foresters to consider, in their use of machinery, chemicals, and planting patterns, the inevitable off-site impacts of land-surface modifications. Simple, educational approaches are appropriate because the 'message' needs to cross professional disciplinary boundaries and be interpreted in the driver's cab of a tractor as well as in the board room. Fig. 5.4(a) and (b) summarizes some of the management guidance available to UK and Canadian foresters respectively.

5.4.2 Regional effects

Newson and Calder (1989) label forest hydrology a 'regional science'; by this they mean that the particular set of intrinsic (forest) and extrinsic (climate) factors which evoke a particular hydrological effect is almost unique within regions. One cannot, therefore, extrapolate the water use by a conifer forest in continental tundra to aid prediction of the effects of groves of eucalyptus in semi-arid India. These process-controlled contrasts are drawn out by Table 5.5.

Furthermore, the *spatial scale* of forests has an effect on their hydrological impact (edge effects often magnify the impact of evaporative processes, see Fig. 2.6); the land cover which they replace will control the *relative impacts*, and wholly local or national forest *management strategies* will complete the geographical variability.

Before this regional variability was formally rec-

ognized by forest hydrologists and incorporated in research design and techniques of extrapolation it had been considered impossible to reach scientific consensus, particularly on the impacts on loss rates by interception and transpiration. Catchment research variously reported more runoff and less runoff from forests in comparison with neighbouring non-forest land covers.

Similarly, controversy raged—and still does—over the effects of forests on flood and drought extremes, sediment yield, pollution, and habitat viability (see, for example, Soutar, 1989; Moffat, 1989). Here site and management characteristics are of particular significance: as one 'descends' the storage cascade which controls catchment runoff there is much less generality of controlling principles than at canopy level where the interception ratio is the result of relatively simple crop height, rainfall pattern, and advected energy properties.

Even at canopy level it is vital to separate transpiration and interception as processes, together with their controls. As an indication of this necessity, Roberts (1983) has tabulated the *transpiration* rates of forests throughout Europe: they are similar because tree species differ little in their physiological need for water (fast-growing species such as eucalyptus are an exception) and, where canopy covers differ, the sub-canopy of shrubs and herbs 'makes up' the atmospheric deficit in humidity (Table 5.6(a)). Thus interception, largely dependent on

Table 5.6(a). Annual transpiration rates of various tree species

Species	Location	Transpiration (mm y^{-1})
Sitka spruce	Slaidburn, Yorks., UK	340
Norway spruce	Germany	362
Norway spruce	Germany	279
Norway spruce	Plynlimon, Powys, UK*	290
		340
		330
Scots pine	Germany	324
Scots pine	Thetford, Norfolk, UK	353
Scots pine	Crowthorne, Berks., UK	427
Oak (sessile)	Germany	327
Oak (sessile)	UK	320
Beech	Belgium	344
MEAN		333

* Annual total given for three separate years

Source: Roberts (1983).

rainfall duration and ventilation, may either exceed transpiration or come a poor second in loss rates (as in dry areas) to transpiration. This separation of processes is an essential element of the simplest models of forest water use (see Newson and Calder, 1989; Calder, 1990).

5.4.3 Water use by forests

Whilst water use applies exclusively to transpiration this process was confused and conflated with interception for so long that we now often apply it to total water losses from a given vegetation canopy. In the previous section we deduced that canopy processes were regionally more predictable than those controlled by soils, geology, or management practice; here we contrast the canopy atmospheric link under tropical and temperate conditions.

Newson and Calder (1989) separate, within the tropical zone, the humid and seasonally dry climates of, respectively, Java and southern India. In the former, soil moisture is seldom limiting to transpiration and there is little advection of energy through the extensive forests (>10 km diameter). In southern India soil moisture depletion and the physiological controls exercised by forest species are much more important.

In the *humid tropics* it has been thought possible to estimate that all net radiation is converted to latent heat (as water vapour); Calder et al. (1986) vindicate this view, recording losses by transpiration of 886 mm and by interception of 595 mm at their Java site. Tree species appear to differ more as a result of their canopy position in the hierarchical

tropical forest than from their water requirements.

By contrast, in southern India and, from analysis of the literature, in similar climates with seasonal moisture stress in the soil, some species of eucalyptus control their water use according to soil conditions and others according to atmospheric conditions; the latter can effectively act as 'water pumps' where the roots are continuously supplied. Greenwood *et al.* (1985) measure annual transpiration from *Eucalyptus globulus* and *Eucalyptus cladocalyx* at 2700 mm in Western Australia (from a rainfall of only 680 mm—the supply to the trees coming from groundwater). George (1990) reports the use of these species to reduce the seepage of salty groundwater, detrimental to arable agriculture, again in Western Australia.

Under *temperate* conditions with an upland climate of cool, wet, and windy conditions, the interception 'tank' takes a controlling influence (Plate 5.2(a) and (b), and Table 5.6(b); Kirby *et al.* 1991), with ratios of interception losses to transpiration of 1.6:1. In eastern England, in a warmer, drier climate the ratio is almost exactly reversed. As a result tall crop (forest)/short crop (grass) catchment losses for the two distinctive climates are: upland 2.1:1, lowland 1:1.1. Veneklaas and Van Ek (1990) find similarities between temperate and tropical *montane* interception rates but the growth of epiphytic (lodging) plants in the tropical canopy is a further complication; interception ratios of 12–18 per cent are reported for the Colombian Andes. The epiphytes create a larger storage capacity for the Andean canopy (c.10 mm: cf. UK <2 mm) but it is less efficiently emptied under tropical conditions;

Table 5.6(b). Interception ratios for conifers in upland Britain

Authors	Date	Location	Trees	Annual rainfall	Interception ratio
Law (extended by Calder, Newson, and Walsh 1982)	1956	Stocks Reservoir	Sitka spruce	1496 mm	.38
Gash and Stewart	1977	Thetford Forest	Scots pine	595 mm	.36
Ford and Deans	1978	Greskine Forest	Sitka spruce	1639 mm	.30
Courtney	1978	Trentabank	Larch		.20
		Reservoir	Sitka spruce	1100 mm	.49
			Scots pine		.52
Gash, Wright, and	1980	Plynlimon	Sitka spruce	2250 mm	.27
Lloyd		Roseisle	Scots pine	650 mm	.42
		Kielder	Sitka spruce	1400 mm	.32
I.T.E.	1986	Plynlimon	Sitka spruce	2250 mm	.19
		Plynlimon	Larch	2250 mm	.24
		Beddgelert	Sitka spruce	2540 mm	.27
		Plynlimon	Sitka spruce	2250 mm	.24
			Larch	2250 mm	.25
Calder	1990	Crinan Canal	Sitka spruce	2000 mm	.39
Pyatt and Anderson	1985	Kershope	Sitka spruce	1435 mm	.62
Hudson, Johnson, and Blackie	1985	Balquhidder	Sitka spruce	2500 mm	.12

stemflow is therefore virtually zero (Fig. 5.5(a)). At the time of writing efforts by forest hydrologists and hydrometeorologists are yielding increasing knowledge of interception rates by tropical-forest canopies (Table 5.6(c)) but these are unlikely to be of long-term benefit without the parallel study of transpiration rates to enable the controlling conditions and rates for both processes to be modelled.

5.4.4 Riparian/phreatophyte effects

Table 5.5 identifies rooting depth and soil permeability as regional factors influencing forest water use. Forest *scale* is also important, via its influence, in advected energy regimes, on ventilation—an important effect on vapour pressure deficit.

A combination of highly favourable circumstances

Plate 5.2. (a) *Opposite* Interception of rainfall on a conifer canopy. (b) Evaporation of intercepted moisture from an upland conifer plantation.

for transpiration occurs when various bands of trees follow river floodplains or wetlands in dry climates; effectively these are the conditions reported for eucalypts in Western Australia, but riparian and phreatophyte forest effects have been studied closely in western USA. Phreatophyte vegetation 'stands with its roots in water' for most of its life and can therefore almost 'pump' water by transpiration; the problems raised by such excessive water use along the Colorado River are such that campaigns have been waged to remove Tamarisk (or Salt Cedar) from the floodplain so as to augment alluvial groundwater. Graf (1985) tables phreatophyte transpirational rates of up to 2800 mm yr^{-1} for tamarisk and 2480 mm yr^{-1} for *cottonwood*, another floodplain occupant; of course precipitation inputs along the Colorado are minimal compared with these rates. However, care should be taken in

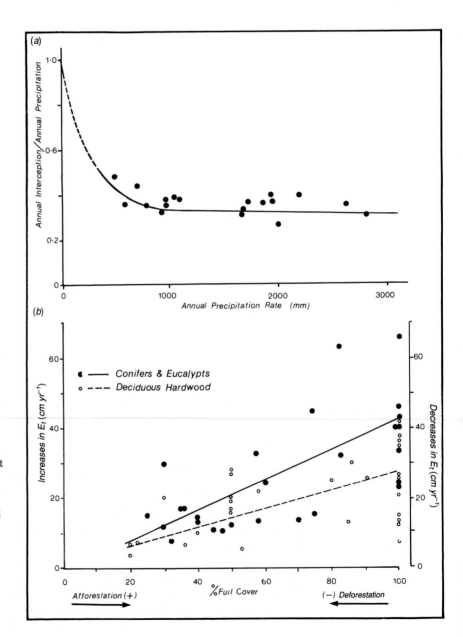

Fig. 5.5. (a) The effect of mature conifer plantations on the interception ratio at different annual precipitation depths (after Calder and Newson, 1979). (b) Results of an international survey of runoff changes when catchments are afforested or clear-felled (after Bosch and Hewlett, 1982).

Table 5.6(c). Throughfall, stemflow, and interception from rainfall in selected tropical forests

Forest type	Location	Percentage of total rainfall		
		Throughfall (T)	Stemflow (S)	Interception (I)
Eucalyptus hybrid	India	80.7	7.7	11.6
Shorea robusta	India	66.4	8.3	25.3
Shorea robusta	India	54.6	7.2	38.2
Pinus roxburghii	India	74.3	3.6	22.1
Alstonia scholaris	India	57.0	17.0	26.0
China fir plantation	Taiwan	91.7	0.8	7.5
Zelkova plantation	Taiwan	90.7	1.4	7.9
Natural hardwoods	Taiwan	85.9	1.7	12.4

Source: Modified from Oyebande (1988).

extrapolating riparian/phreatophyte forest hydrology to any 'normal' conditions.

5.4.5 Predicting forest influences on runoff volume

We have a scientific paradox in land-use hydrology: the need for *prediction*, but the regional *uniqueness* of the results. This paradox is a central motivation to much of the measurement and modelling activity in forest hydrology. The simplest process to measure extensively is interception loss below forest canopies (see Chapter 2) but the resulting predictions, such as the Calder–Newson equation below are limited in their application to those conditions where conifers grow in close proximity in a cool, wet, windy climate (effectively the country of origin). Calder and Newson (1979) put forward the simple prediction for total loss from a forested area:

$$E_{tot} = E_t + f(P.\alpha - w.E_t)$$

where: f is the fraction of the catchment with forest canopy, P the annual precipitation; α the interception ratio (see Fig. 5.5(a) and Table 5.6(b)); and w the fraction of the year when the canopy is wet. They later had to introduce modifications for shorter periods and for comparisons with intermediate height crops like heather.

Bosch and Hewlett (1982) make an ambitious compilation of the results of many catchment experiments worldwide to plot the forest effect, largely from felling operations which increase streamflow yields. Fig. 5.5(b) shows how runoff increases slightly more for conifers and eucalypts than for hardwood species after felling (and vice versa) but one should note that these results are often gathered from different climates as well as different species.

5.5 Agriculture and hydrology

Plantation forestry may be considered as trees grown by agricultural principles and whilst, strictly speaking, agriculture deals with shorter crops without the evapotranspiration 'advantages' of trees, we may expect similar impacts in terms of land management. There are, however, some notable constraints with forests:

• Agriculture and forestry are often allocated to different climatic or soil zones; earlier remarks about regional effects may therefore prevent comparisons.
• Agriculture has a *shorter crop cycle* with a relatively minor impact in establishment and harvesting, compared with those of forestry.
• Agriculture (and especially horticulture) makes considerable use of *irrigation*, of *chemicals* to control growth and pests, and of modified ground surfaces such as *terraces, mulches*, etc.

Not surprisingly, Arnell's review (1989) of land-use hydrology across eighteen nations reveals that two categories of impacts dominate: vegetation change and fertilizer application (with studies in fifteen and thirteen countries, respectively). Whilst we treat the non-point pollutants entering streams from agriculture in Chapter 8 we need here to investigate the physical impact of changes from natural land surfaces to agriculture and within agriculture in terms of improvements to drainage and seasonal vegetation changes.

5.5.1 Cultivation effects

In Chapter 3 we discussed the hydrological and hydraulic properties of 'smooth' and 'rough' surfaces in terms of surface runoff. There are other

Plate 5.3. Ground preparation for upland conifer forest.

effects as Fig. 5.2 suggests—cultivation increases roughness but also increases depression storage and infiltration capacity. In changes from a pristine vegetation surface, where this has both sheltered and bound the soil, cultivation is a major physical change both at the surface in terms of atmospheric relationships and at depth (see Plate 5.3). Cultivation has a finite depth (affecting soils in what is known as the *plough layer* and possibly smearing or compacting them below) and it has an orientation with respect to runoff routes (parallel or normal to slopes).

Reid (1979) made measurements of the *surface roughness* of ploughed arable land in lowland England; he showed how the capacity for depression storage decreased from 6 mm to just over 3 mm during a winter of rain, frost, and other forces of compaction. Infiltration rates were not measured but other studies have reported reductions under wheel-tracks from heavy farm vehicles; tracking is also implicated in erosion of soils. However the fine tilths prepared for autumn seed-beds in the UK have also been blamed for the 'capping' that forms the impermeable surface layer, promoting surface runoff in winter (Speirs and Frost, 1985). Careful choice of cultivation techniques is of even greater hydrological importance under tropical conditions, especially where forests are cleared (Reynolds, 1990).

5.5.2 Drainage effects

Section 5.3 and Fig. 5.2 have already hinted at the major problem of elucidating the effects of drainage

on river flows: effective drainage should increase soil *storage* for precipitation as well as conduct the *outflows* from the soil efficiently to a channel—these may have equal effects on the volume and timing of runoff. If storage is increased (effectively drying the soil between precipitation events to aid better crop growth) it may well be that an affected stream will experience a later and lower runoff peak than a nearby watercourse receiving runoff from undrained land. On the other hand, where the soil does not drain easily between precipitation events much more of each event enters the drain system—quickly in the case of open surface drains but via slower surface routes where there is underdrainage (the route to the drains being effectively saturated).

The complicating factors are so numerous that controversy has long surrounded the agricultural (and forestry) drainage impact, especially where the impact of increased or reduced floods (or droughts!) is predicted for sites far downstream from the drained area. Nevertheless, the drainage of individual plots and fields, if part of a national policy, can gross up into a major land-management impact in quite large catchments (Higgs, 1987; Robinson and Armstrong, 1988).

It is possible that drainage studies rival forest hydrology in qualifying as a 'regional science'; some authors discover results suggesting opposite impact between sites (see Newson and Robinson, 1983, versus Robinson and Newson, 1986—and Fig. 5.6) whilst others have proved opposite effects on the same site depending on season and on the scale at which measurements were made, i.e. field or catchment (Robinson and Beven, 1983; Reid and Parkinson, 1984; and Fig. 5.7).

Fortunately Robinson (1990) has recently gathered many of these apparently conflicting studies between two covers and Table 5.7 shows the impact of pipe underdrainage schemes (of the type shown in Fig. 3.9) on the parameters of the unit hydrograph at the field scale. He goes on to investigate the characteristics of the experimental plots in detail, showing that drainage of deeper, more permeable soils tends to increase storm-discharge peaks; in shallow, dry soils there is a decrease. In the latter situation there has clearly been a change from surface to subsurface flow routes as a result of effective drainage.

Robinson's other major point is that at the catchment scale, the effects of other aspects of agricultural improvement such as channel excavations, may reveal a tendency for flood peaks to increase as a result of the farm drainage (the link is however

Fig. 5.6. The impact of forest drainage (open ditches) in relation to agricultural drainage (pipes) on the unit hydrograph of drain outfalls. At Tylwch agricultural drainage reduced peak flows (Newson and Robinson, 1983); at Coalburn forest drainage increased peak flows but a recovery followed (Robinson, 1986); the results for Tanllwyth, Aberbiga, Upper Severn, indicate the variability of the drainage impact in mature forests on different soils (Robinson and Newson, 1986).

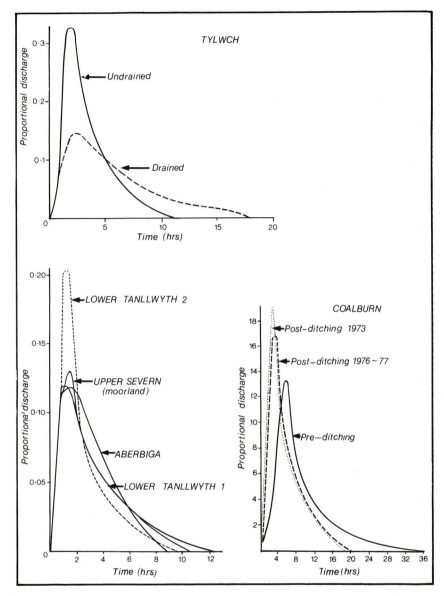

indirect). Figs. 5.6 and 5.7 include both plot scale and catchment-scale results for this reason.

5.5.3 Canopy effects

Agricultural-crop canopies are shorter and often less dense than those of either natural or plantation forests. Processes on the canopy do not differ; interception and transpiration both contribute to crop water use—typical values of crop-interception coefficients are shown in Table 5.8. Once again, however, we are in a regional science, with man-agement, soil, and climate variables conspiring to corrupt the results of general predictions; in an era of climate change (including changes in stomatal functioning because of increased carbon dioxide) this lack of predictability is serious.

Under present conditions of relatively abundant water supplies (either natural or enhanced by reservoir supply schemes) crop water use is dominated by transpiration under irrigation-water deployment but, in future, canopy processes in crops need to be understood in order for *spray irrigation* to be applied with less waste by interception. Surface-water

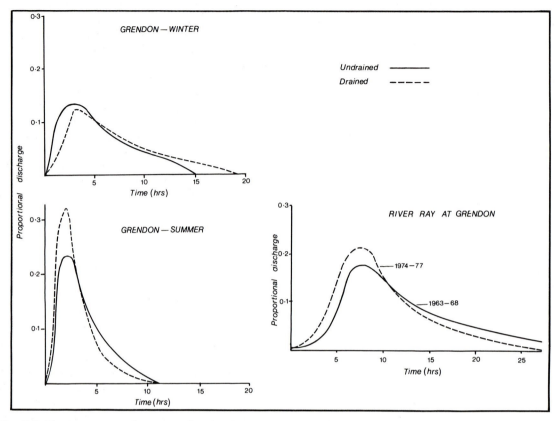

Fig. 5.7. The importance of season and scale of measurement on the unit hydrograph at Grendon Underwood. The drained plot yields higher (summer) and lower (winter) peaks than the undrained plot because of the importance of cracking of the clay soil in transmitting rainfall to drains. For the catchment as a whole (River Ray) the evidence of the main drainage period of the 1970s is that peak flows are increased (after Robinson, 1990).

Table 5.7. Comparison of 1-hour average unit hydrographs, for drained and undrained land, described by the time to peak (Tp) in hours, and the peak hourly flow (Qp) as a proportion of the total storm flow

Site	Undrained		Drained	
	Tp (hours)	Qp	Tp (hours)	Qp
Tylwch	2.0	.32	2.5*	.13*
Staylittle	2.0	.23	2.5	.14*
Withernwick	5.0	.09	4.0	.18*
Grendon				
(winter)	3.0	.13	3.5	.11
(summer)	2.0	.23	2.0	.31*
Blacklaw Moss	6.0	.07	2.0*	.18*
Summary parameters†				
Mean	3.6	.168	2.9	.148
Standard deviation	1.8	.105	0.8	.031
Coefficient of				
variation (%)	50	62	28	21

* Significantly different from undrained land at .05 level.
† Excluding Grendon summer events.
Source: Robinson (1990).

Table 5.8. Interception by agricultural crops

Crop	Plants m^{-2}	% intercepted R = 175 mm 1–31 July	% intercepted R = 65 mm 1–25 Sept.
Maize	9	17	50
	16	20	55
	25	19	60
	36	32	77
Soya beans	9	0	28
	16	7	31
	25	12	56
	36	17	47
Oats	16	11	5
	25	10	6
	36	10	13
	49	15	26
Vetches	25	12	32
Beans	25	25	35
Lupins	25	22	58
Peas	25	10	—
Clover	Broadcast	40	—

Source: Penman (1963).

applications which avoid the canopy should also be planned on the basis of improved climatic predictions of total *crop water use*, of at least a weekly periodicity.

Crop-canopy cover is also important in controlling surface *soil loss*; one of the changes in agriculture in the UK charged with increasing soil erosion is the practice of autumn-sown arable cropping which leaves a friable soil unprotected by a canopy all winter (Boardman, 1988).

5.6 Urbanization

One of the salient phases in the development process in the modern world is the wholesale movement of people from rural homes and livelihoods to urban centres. Even in the case of Britain two centuries ago, during the Industrial Revolution, this movement was extraordinarily rapid, the switch from an agrarian to an urban and industrial society occurring in a period of under a century.

The corollaries of urban growth were, and continue to be:

• Hazardous conditions in respect of *water supply* for both health and fire within the densely-settled parts of cities (especially the poorer parts).
• Problems of *waste disposal* and a surplus of drainage water gathering from impervious surfaces.

Often the waste and drainage problems are cured first with an obvious disposal route in most sites being a joint one (with the drainage water propelling the waste through buried pipes) to the nearest watercourse. Since the comprehensive supply of clean water to the urban population followed later, there is an interim (hazardous) condition in which humans use grossly polluted rivers for cleansing themselves and even for securing water supplies. The completed urban development can amount to a very extensive and very profound modification to natural climatological and hydrological processes: one should not merely consider urban effects as operating through the change brought about to surfaces.

5.6.1 Elements of urban surfaces

The section through a compressed slice of urban and suburban surfaces shown in Fig. 5.8(a) indicates that there is not one surface but many; there are also many routes for runoff. The original concept of urban surfaces was both far too unified and far too emphatic about the *impermeability* (and hence high rates of runoff) to be anticipated by contrast with rural surfaces. Hydrologists have now realized that whilst many of the early municipal hydrological design problems were posed by large, 'hard', developments the total urban patchwork of surfaces does not yield 100 per cent runoff from each storm. Some characteristic rates are shown in Table 5.9. Clearly pro-active urban planning can and should make maximum use of hydrological information such as this in controlling the rate, location and mix of urban developments.

5.6.2 Urban drainage

There are two functions of urban drainage: *storm drainage* which removes surplus rainfall (otherwise constituting a flood hazard and dangerous transport conditions) and *sewerage*, the provision of routes for waterborne wastes created by washing and toilet facilities. It is now common in the developed world to separate these two forms of drainage; however joint systems prevail in many countries with storm water an integral part of the functioning of sewerage. The unacceptable effect of such joint systems (see Chapter 8) is that in storms greater than a certain (low) threshold size, sewers spill their contents to the nearest natural watercourse, polluting it in the interests of avoiding spillage on to the streets or damage to sewage treatment facilities.

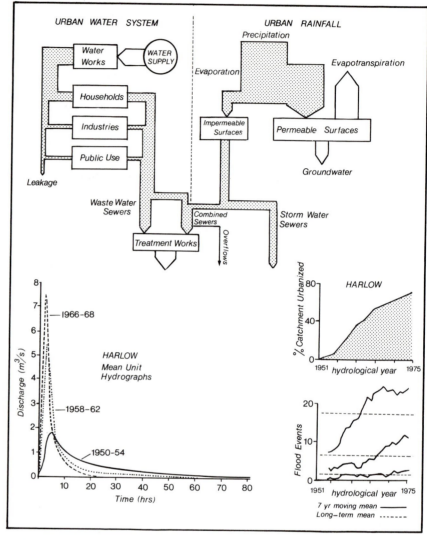

Fig. 5.8. The urban runoff system analysed systematically, (after Lindh, 1990), and (below) its impact on flood hydrographs (after Hollis, 1974) and flows over various thresholds as urbanization has continued (after Woolhouse, 1989).

As with forestry and agriculture, local custom and practice is essential knowledge when considering hydrological impacts on receiving streams. The urban hydrologist must first map the true catchment and 'stream' network as revealed by the data kept by the city engineer. The network of storm drains and/or sewers must follow the given pattern of topography and development: urban-drainage design does not precede urbanization but follows it. As a consequence, some parts of the network are of shallow gradient and its contents may require pumping; other sections are steep and may regularly overflow in heavy rain, especially if new developments have been 'bolted on' to the upstream end of an already stretched pipework capacity.

Whilst storm-drain intakes are generally protected against the intake of large street-level material it is common for *debris* and *sediments* (Ellis, 1979) to block drains; the resulting floods are doubly troublesome since they contain high levels of road-derived pollution and sewage.

In recent years the Victorian hangover in urban drainage (British techniques permeated the world) has been overthrown in favour of brand-new structural approaches including *storages* and *brakes* to avoid the rapid runoff and intermediate flooding/ sewer overflows characteristic of older techniques. Fig. 5.9 shows commercially available techniques of drainage to serve new urban developments in the UK.

Clearly, too, properly planned urban developments can:

- maximize surface storage
- maximize infiltration to sustain aquifer recharge
- minimize the extent of uninterrupted solid surfaces
- maximise use of innovative, less impermeable, materials
- separate flow routes from sources of pollutants.

5.6.3 Experimental evidence at various scales

Urban hydrology is not an easy or pleasant occupation: vandalism besets instrument systems, there are dangers from traffic and property owners, and the quality of water one measures is often equally, if not more dangerous. The early experimental studies of urban runoff, therefore, tended to be 'black box' studies of change from rural land cover during urbanization outside the urbanizing zone itself. Fig. 5.10(b) and (c) illustrates changes in the unit hydrograph and in flow frequencies in the Canon's Brook, Essex, UK during three decades of urbanization. Eventually, however, it has been found essential to look at processes in detail, quite literally at street level.

The black box techniques for predicting the effects of urbanization on runoff, especially on flood peaks, tended to use a simple *rational formula* linking the flood discharge to catchment area, e.g.:

Table 5.9. Approximate microclimatological modifications as a result of urbanization

Element	Comparison with rural environment
Contaminants	
condensation nuclei and particles	10 times more
gaseous emissions	5–25 times more
Cloudiness	
cover	5–10% more
fog, winter	100% more
fog, summer	30% more
Precipitation	
totals	5–10% more
days with less than 5 mm	10% more
snowfall	5% more
Radiation	
global	15–20% less
ultraviolet, winter	30% less
ultraviolet, summer	5% less
sunshine duration	5–15% less
Relative humidity	
winter	2% less
summer	8% less
Temperature	
annual mean	½–1°C more
winter minima (av.)	1–2°C more
Wind speed	
annual mean	20–30% less
extreme gusts	10–20% less
calms	5–20% more

Source: Landsberg (1970).

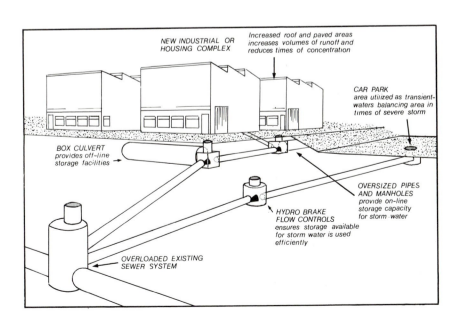

Fig. 5.9. The potential for using the underground system of storm sewers to 'break' and store rapid urban runoff before discharge to streams.

$$Q\ (\text{m}^3\ \text{s}^{-1}) = \frac{A \times I}{3600}$$

where Q is peak discharge, A is catchment area (ha) and I is the rainfall intensity (mm h^{-1}) (for a full treatment see Shaw, 1983). However, this almost total conversion of rainfall to runoff is inaccurate; relationships involving the timing of flood runoff to area continue however, to be substantiated empirically (see Fig. 5.10; and Leopold, 1991). Flood-peak relationships with catchment area tend now to be specific to a single development or include as a parameter the proportion of the catchment with storm drains. As with agricultural drainage effects in Section 5.5.2 we must be clear in urban hydrology whether the results we read are gathered at the *street* (cf. plot or field) level or at the larger unit of the urbanized/drained *catchment*.

Whereas the first long-running hydrological studies of urbanizing catchments reported spectacular increases in flood peaks (e.g. Fig. 5.10(b); and Hollis, 1974) it has generally been found that the urban influence is not of the same degree for all surfaces (see Table 5.10) and at all flood-return periods (see Hollis, 1975). Hollis has continued the search for detail in the urban runoff process in the UK. Hollis and Ovenden (1988a) revealed an interesting seasonal effect caused by frost action on paved surfaces: winter *infiltration rates* exceed those of summer by a factor of 3.2, the reverse of what might be expected in rural areas. The runoff percentage for the larger storms was by no means 100 per cent—it seldom exceeded 10 per cent. Going on to studies of other typical surfaces, Hollis and Ovenden (1988b) revealed much higher *runoff*

Fig. 5.10. Empirical guidance to predicting the hydrological impacts of urbanization. (a) The effect on return periods of given flows (after Leopold, 1968). (b) A similar approach using ratios of increase in mean annual flood but incorporating the effects of both impervious surfaces and drainage (after Packman, 1979). (c) The effect of urbanization (impervious surface) on the lag time of flood response (after Leopold, 1991).

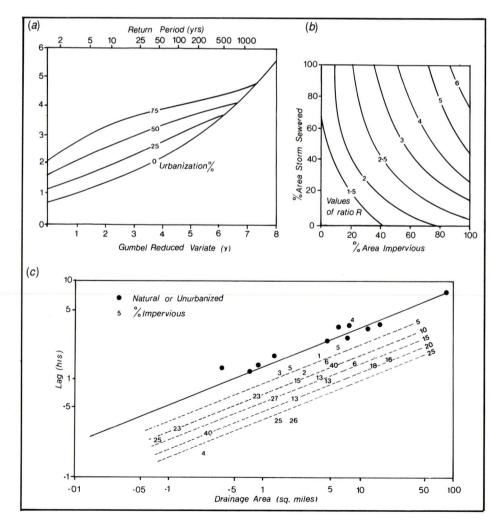

Table 5.10. Runoff rates from urban surfaces

Description	Runoff coefficient	
	Min.	Max.
Pavement, asphalt or concrete	0.80	0.95
Gravel roads and shoulders	0.40	0.60
Roofs	0.70	0.95
Commercial		
downtown	0.70	0.95
neighbourhood	0.50	0.70
Industrial		
light	0.50	0.80
heavy	0.60	0.90
Residential		
single family urban	0.30	0.50
multiple, detached	0.40	0.60
multiple, attached	0.60	0.75
suburban	0.25	0.40
Apartments	0.50	0.70
Parks, cemeteries	0.10	0.25
Playgrounds (unpaved)	0.20	0.35
Railroad yards	0.20	0.35
Unimproved	0.10	0.30

coefficients for roofs (average 57 per cent, varying from 28 to 90 per cent). Rainfall intensity proved to be the variable with most predictive power. Depression storage amounted to under 1 mm.

Clearly further calibration and monitoring efforts are required for both urbanizing and adjacent rural catchments at all scales (plot, car park, street, drain) and for long durations so as to experience the full range of extremes. It is likely, however, that water-quality considerations will play a major part in the reform of urban drainage (see Chapter 8).

5.6.4 Effects on urban river channels

In many urban areas the former natural-surface channels have been culverted underground over long distances or so completely constrained between concrete banks and beds that they are effectively open drains. In Britain much of this loss occurred during the Victorian period of urban expansion; surface streams were either an obstruction to transport routes or the site for dumping of excrement and other wastes.

Those open non-rigid channels which remain in or close below urban areas are in receipt of a changed inflow regime in terms of flow, sediments, and chemistry. Because channels adjust to these variables (the chemical effect can be via vegetation) it is not surprising that such channels change

rapidly. Empirical studies of channel dimensions reveal an erosional problem, largely the result of increased flood flows (e.g. Park, 1977; Knight, 1979). The latter study recorded cross-sectional capacity increases in the order of sixfold. Roberts (1989) has advanced evidence from a comparative study of UK urban channels that inpermeable catchments, with channels adjusted to more frequent bankfull flows are most susceptible to the impact of urbanization on channel capacity.

5.7 Land-use hydrology: what next?

We have learned so far that the investigation of land-use effects on the hydrological cycle is a major stimulus to research; however, we have also discovered that unambiguous and widely extrapolatable results are hard to come by. Is there sufficient consensus and perception for river-management institutions, both national and international, to seek control over land-use change and land-management practice?

5.7.1 International practice

We have already referred to the complex path by which scientific knowledge (especially uncertain knowledge) becomes incorporated in institutional practice. Not surprisingly, therefore, it is not common to find national water-management bodies with significant powers over land use and land management. In the UK, largely stimulated by European Community legislation, the influence of the water interest is extending slowly (see Newson, 1991); internationally, the picture is one of government institutions attempting co-ordinated resource planning of land and water within a river-basin framework (Newson, 1992) but with water-quality worries spurring on the change (e.g. Canada, New Zealand). Pollutants, or their effects, are visible and their sources often clear; in policy terms, therefore, pollution problems are easier to react to. Mere hydrological changes are often imperceptible or may be compounded by climate change.

Nevertheless, as the science of hydrology develops further we are almost certain to see a positive intervention by all nations to prevent the negative effects of the use of land, and imperfections in our manipulation of water supplies from jeopardizing future world development. L'vovitch and White (1991) list the main direct actions as:

manipulating stream channels
damming watercourses
draining wetlands
transferring water
extracting groundwater
irrigating agricultural lands

Chief amongst the indirect actions are:
spread of cities
ploughing of grasslands
felling of forests

The authors go on:

Thereby the hydrologic cycle and associated ecosystems were disturbed in the search for agricultural and industrial production and urban amenities. Determining the full nature of those transformations is handicapped by lack of statistics as to the human activities and waterflows, and also by unprecise understanding of how natural systems have been affected. (p. 235)

Despite these areas of ignorance, many international guidance documents now advocate a precautionary approach to river-basin development: UNEP/UNESCO (1990) warn that:

The environmentally sound management of water resources projects should be implemented by an active interdisciplinary and intersectoral learning process. The interaction between water projects and their host river basins should be studied in depth.

OECD (1989) state that,

for water management the river basin or catchment area often is advocated as the most appropriate spatial unit for several reasons which include emphasizing the interdependency of water quantity and water quality, water and adjacent land-based resources and upstream and downstream effects.

5.7.2 River basin land-use data

Given the hindsight of 10 000 years or more in their review of Holocene river dynamics in the UK, Macklin and Lewin (forthcoming) conclude that changes wrought in the valley/channel system over such time-scales are 'climatically-driven but culturally blurred', in other words the effects of land use take second place to those of climate. This is useful guidance but one must consider the danger that it, too, is a regionally valid conclusion without general applicability. Furthermore, the impact of human culture is now much more widespread than at any other stage of the Holocene.

Given, therefore, that international practice is increasingly validating the relationship between catchment activity and streamflow/ecosystem response it is essential to consider data-collection systems which allow relevant extension of hydrological information into land-use information. Such extensions can be made by using:

• GIS (geographical information systems) to store and manipulate the two types of information pooled from existing sources.
• Satellite surveys to update land-use change in real time, including seasonal changes of, say, cultivation (Gregoire, 1991).
• The development of interventionist agricultural policies including production quotas and buffer zones will provide information akin to that held by planning authorities on urban land-use patterns.

The next thirty years will see the interesting if slightly neurotic interplay of climate change, existing land use and land-use response to climate change in determining river-management scenarios.

5.8 Effects deriving from water management

Land-use effects on the basin hydrological cycle are indirect on the channel component, by comparison, that is, with the profound manipulations of flow brought about by *impoundment* and *transfers*, *channelization* and *regulation*. The latter term has achieved a general value in describing all the more direct human interventions which control flow regime (and often quality and biotic status) of rivers world-wide.

5.8.1 Water abstraction and use

We have already outlined the relatively small proportion of global surface, fresh water which is utilized approximately 10 per cent or 3600 km^3 per year.

Locally, however, and in the context of some large and internationally important rivers, water abstraction and consumptive use threatens both downstream supply and the river ecosystem. Fig. 5.11(a) illustrates the situation in the Colorado River, a system with other management problems (see Sections 4.5.4 and 5.4.4). A similar situation is potentially developing on the Nile, where nine African countries compete to dam and abstract the water of the river with decreasing residual flows entering Egypt's prestigious Aswan Dam (Schwarz *et al.* 1991).

Not only does the problem of abstraction affect

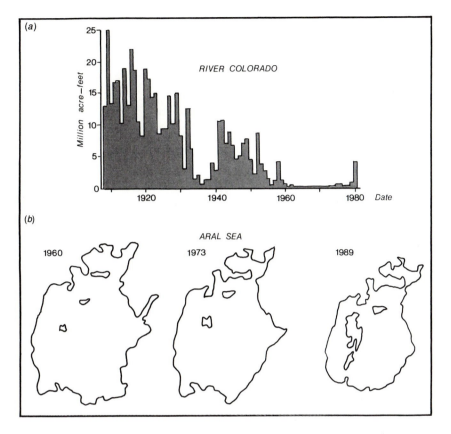

Fig. 5.11. Regulated rivers and water use in (a) the Western USA, and (b) South Central Asia (after Williams and Aladin, 1991). The Colorado graph shows surplus residual flow in the river after supplies are satisfied. The area of the Aral Sea reflects upstream use of the inflowing streams for irrigation.

the available flow resource in a river via a consumptive (evaporative) use; in the case of human use the return flow consists of sewage, and where industry 'uses' the water it is often returned to the river either polluted or heated to non-natural temperatures. In some cases it may be abstracted from one catchment but returned to another (e.g. in the UK the long-distance transfers from the Welsh hills to Liverpool, Birmingham, and other distant centres).

L'vovitch and White (1991) estimate the growth of mankind's water manipulations; historically, after 1951 three times as many large reservoirs (>100 million m^3 storage) have been built than in all previous years (Fig. 5.12(a)). The surface area of water in all the large reservoirs is comparable with that of France. Hydropower comprises 23 per cent of all power outputs. Aquifers are still being depleted, by up to 3 m in the High Plains of the USA. The authors correlate the loss of floodplains and wetlands with the increase in damaging floods (see Chapter 4). Their conclusion is that the new agenda for human water use is to cleanse our

return water. Waste reduction, waste treatment, and water-use efficiency are fundamental to the continued exploitation of the basic resource (see Fig. 5.12(b)).

5.8.2 Irrigation

Section 4.5.4 charted the spread of irrigated agriculture and some of its hydrological pitfalls both within the irrigated area and downstream.

Irrigation makes very heavy demands on water resources; in Egypt the national dependence on irrigated agriculture is almost total, resulting in the distribution of water use shown in Table 5.11. Including drainage flows and seepage from the system nearly 88 per cent of the nation's available water is used in irrigation. Further works on the Nile may yield sufficient for another ten years' population growth (Higgins et al. 1988). Cereal yields of 1.5 kg m^{-3} of water supplied are practically feasible but few nations have reached one third of this optimum.

Perhaps the most widely publicized hydrological

Fig. 5.12. The growth of both number and volume of reservoirs by continent and the predicted pattern of future global water use (after L'vovitch and White, 1991).

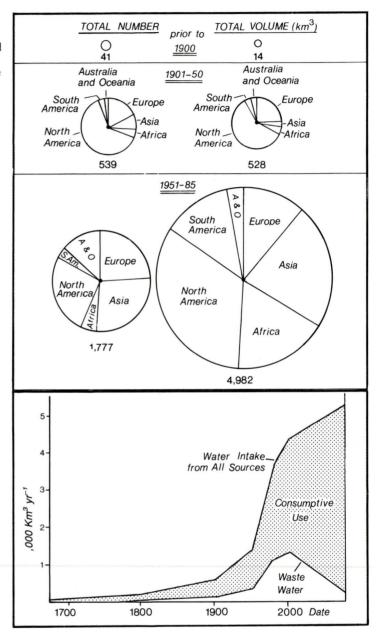

impact of irrigation development in recent years has been in central southern Asia where Soviet plans to irrigate up to 9 million ha. have virtually destroyed the Aral Sea, whose level fell 8 m in twenty years (Fig. 5.11(b)). It is predicted that in the current decade the Aral Sea will have seen its natural inflow of 54 km³ yr⁻¹ decrease by 51 km³ yr⁻¹. Former Soviet plans to divert north-flowing rivers into the crisis zone were scrapped in 1985.

5.8.3 Land drainage/flood protection and 'channelization'

The natural river channel is the outcome of water flows and sediment loads of a certain magnitude and frequency (see Section 3.6.2). It is tempting to 'improve on nature', particularly in terms of

- The *capacity* of a channel to conduct floods of a greater magnitude (options are illustrated in Table 5.12).

Table 5.11. Estimates of the annual surface water budget in Egypt, based on outflow from High Aswan Dam

Use	%
Consumptive use in irrigation	54.4
Drainage to sea	29.6
Navigation and pollution control	7.0
Canal seepage to water-table	3.7
Municipal	3.3
Industrial	0.6
Other	1.3

Source: Lenton (1985).

- The *stability* of river-channel beds and banks in order to reduce erosion of productive land, boundary changes and the undermining of river structures.

The overall process of channel 'improvements' (an anthropocentric term in this case) is now commonly known by its American title of channelization (Brookes, 1988, and Table 5.12). The emphasis of its purpose in England and Wales has changed recently from *land drainage* to *flood protection*, largely because of the fall from political grace of intensive agricultural production with the arrival of food surpluses. It had also been suspected by hydrologists that these two activities were incompatible—in fact that land drainage intensified floods.

A review by Robinson (1991) showed that for four British catchments the effect of arterial drainage (effectively channelization) was to shorten response times to rainfall and increase peak flows; whilst local flooding was reduced the problem was moved downstream in two of the catchments. Robinson also recorded an increase in minimum flows after the schemes were completed; whilst annual maximum discharges increased by 14–88 per cent, minima increased by 10–57 per cent, the result of drawing down floodplain aquifers (a phenomenon measured in Ireland by Essery and Wilcock, 1990).

The impact of channelization depends upon the extent of the scheme which can be defined quantitatively, in terms of the flood-peak flow which the new channel is designed to carry (e.g. 10-year—cf. the adjustment of natural channels to a 1- to 2-year flood) or qualitatively in terms of operations (Table 5.12).

Fig. 5.13 shows that the impacts of channelization, like those of regulation, extend well beyond the change in channel flows. Impacts are so widely reported in such negative terms by environmentalists that restoration schemes are now commonplace, calling on the expertise of geomorphologists to reinstate meanders, bars, riffles, pools, and other natural features (see Gore, 1985).

5.8.4 Reservoir operation: direct regulation of river flow

Whole books are now available on this subject; the number of dams across world rivers continues to grow despite popular protest and scientific concern over impacts downstream. A large number of research projects have been carried out on the effects of river impoundment on flows, channels, pollution, and biota; the experimental context is attractive, with good data available, a measurable

Table 5.12. Terminologies for the methods of channelization

American term	British equivalent	Procedure
Widening Deepening	Resectioning Resectioning	Manipulating width and/or depth variable to increase the channel capacity
Straightening	Realigning	Steepening the gradient to increase the flow velocity
Levee construction	Embanking	Confining floodwaters by raising the height of the channel banks
Bank stabilization	Bank protection	Use of structures such as gabions and steel piles to control bank erosion
Clearing and snagging	Pioneer tree clearance Weed control Dredging of silt Clearing trash from urban areas	Decreasing the hydraulic resistance and increasing the flow velocity by removing obstructions

Sources: Nunnally and Keller (1979) for American terms; Thorn (1966) for British equivalents.

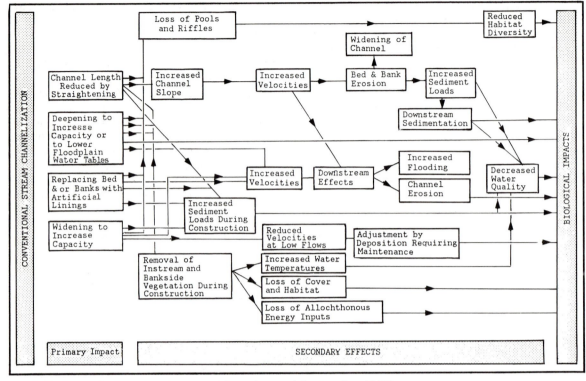

Fig. 5.13. Principal effects of major types of channelization (after Brookes, 1988).

change from natural conditions, and the rare element of flows controlled at the touch of a switch (at least in the reaches nearest to dams). The review by Petts (1984) is very comprehensive, but more recently the challenge has been taken up to agglomerate small-scale studies to national impact assessments (Petts, 1989) and to extend into the influences of impoundments on the world's larger rivers (e.g. Van de Ven *et al.*, 1991).

Petts provides basic conceptual guidance on the impacts of impoundments, presented here as Fig. 5.14. At the hourly/diurnal scale reservoir releases are bound to fluctuate since human demands do likewise, whether for consumption, industrial processes, navigation, irrigation, or power generation. Over longer time periods the storage provided by the reservoir 'creams off' flood peaks and permits the release of augmented low flows. For more than a century, little impact on rivers was thought to result from these changed patterns of flow except some troublesome scouring of fine sediments immediately downstream of dams. Through the work of Petts and others, however, we are able not only to summarize our observations of existing impacts (as

in Fig. 5.14) but to set up matrices or network guides to statutory Environmental Impact Assessment before dams are constructed in future.

Furthermore, we have become increasingly aware of longer-term repercussions occurring in a chain of causation—first-, second-, and third-order impacts. We referred to the need to be realistic about threshold changes and recovery in relation to forest impacts (Section 5.3); in the case of reservoir impacts it has proved possible to date the onset of and recovery from change as shown in Fig. 5.14(b)).

In the larger-scale assessments now being attempted it is common to incorporate all sources of river regulation; for example in Britain, rivers experience impacts from transfers to and from groundwater (e.g. mine pumping), from hydro-electric power generation, from irrigation (often a diurnal pattern) and from sewage-works returns as well as 'normal' reservoir regulations. Box 5.1 indicates the complications which these manipulations introduce to the use of gauging-station flow records.

In terms of regulation effects in much larger basins, Piper *et al.* (1991) suggest that headwater reservoirs have a limited impact on flows in the Mekong; nevertheless, with a catchment area of 32 675 km^2 feeding existing tributary dams the effect on mean annual runoff and mean annual minimum flows can be discerned at a catchment area of 545 000 km^2. A reduction of runoff of 8 per cent at this scale, the result of evaporative losses from the reservoir surfaces, is very serious in terms of future developments.

In Europe the river receiving the most attention on river regulation (as a result of political differences between the twelve nations included in its vast catchment) is the Danube. Since Roman times attempts have been made to improve the river's navigation and to link it by canal with the Rhine (Boucher, 1990); attention has now also turned to water resources and power generation, especially in Hungary and Czechoslovakia. The proposed Gabcikovo-Nagymaros scheme has been the subject of modelling to predict the changes of flood and low flow regime which will follow its completion (Szolgay, 1991). However, the environmental impacts of large river level fluctuations over a 100 km reach, the impact on extreme floods which may 'back up' from the reservoirs, and the costs of re-sewering up to one hundred settlements threaten the completion of the scheme.

Not surprisingly, dam building has attracted its

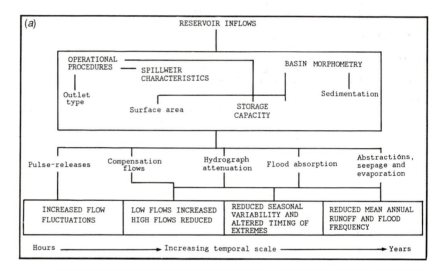

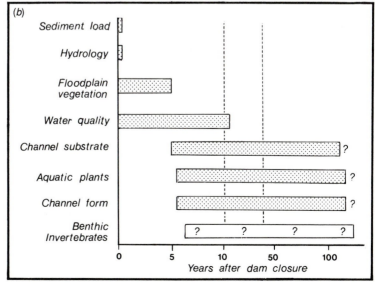

Fig. 5.14. (a) Factors affecting the hydrological characteristics of dammed rivers (after Petts, 1984). (b) Timing of impacts after completion (closure) of dam.

Box 5.1. *UK gauging-stations: regulation effects*

The Register of Gauging-Stations (DOE, 1982) includes data for 1310 stations for the period of record to 1976. Only 11 per cent are classified as 'natural', that is, having no abstractions or discharges, or the variation due to them is considered so limited that the gauged discharge is within 10 per cent of the natural discharge at or in excess of the 95 percentile flow. Three hundred and seventy one stations are unclassified and 793 are classified as affected by flow manipulations, abstractions, and discharges. The following statistics, referring to all stations, reflect the nature of flow regulation in the UK:

• 9.5 per cent are affected by direct regulation, that is, under certain flow conditions the river will be augmented by water from surface water and/or groundwater storage upstream of the gauging station;

• 2.4 per cent are affected by indirect regulation to suit the need for hydroelectric power generation;
• 25.9 per cent are affected by indirect regulation by impounding or storage reservoirs situated in and supplied from the catchment above the gauging station;
• 39.3 per cent are affected by interbasin transfers for water supply, either by abstractions from a reservoir or river intake (25.9 per cent) or by effluent return via sewage works or industrial works (26.6 per cent).

Source: Petts (1988).

---

own protest movement in recent years, with environmentalists' concerns running high over both the downstream impacts on hydrology and the effects on local people and ecosystems. Such movements in Brazil and India are described, respectively by Alvares and Billorey (1988), and Cummings (1990).

Part II: The Freshwater Environment

6. Link section: Elements of freshwater habitat

One major contemporary axiom—the need to conserve biodiversity—forces us to make increasing links between the processes of catchment runoff and those which make the freshwater environment suitable or unsuitable as a *habitat* for life. It makes little sense to separate, within the term 'life', those plants and animals which are truly aquatic, those which spend part of their existence in water and those, like *Homo sapiens*, which enjoy water and depend on water supplies (gained temporarily from water bodies and returned unclean). The principles of ecology state that these variations of organic form and physiology are mutually interdependent and should be treated as such; the principles of resource management suggest that any damage to organisms totally dependent on the freshwater environment is a powerful warning to those less dependent—a repeat of the old 'miner's canary' danger warning. *Pollution control* and *nature conservation* are, in this respect, on convergent paths. If we add recreation by human beings in, on, or adjacent to water, there is a powerful (quantitative and qualitative) pressure for treating rivers and lakes as the outcome of our stewardship of land and water resources.

Part II of this volume encourages the reader to make significant additions to the knowledge commonly grouped under the heading hydrology, additions made from geomorphology, environmental chemistry, and biology. For the complete picture of the environmental approach to river-basin management one needs to understand how runoff, when it reaches wetlands, open waters, and the river channel contributes to a physical and chemical habitat for a very wide range of species, including ourselves. So little work has been done on the linkages between hydrology and the river environment that it pays to ask, as in Chapter 7, some fundamental (even banal) questions such as: 'What is a river?' From that lowly beginning we can assemble a model which brings together the routes for runoff and their chemical and biological signatures together with volumes and timings which control the *average conditions* of habitat (and their very important *extremes*). Such a viewpoint allows a much quicker and more practical appreciation of threats to river environments from, say, pollution or drainage. It also, if translated carefully into practice, permits sustainable management of whole basin ecosystems.

The elements of the freshwater habitat with which we are most concerned are:

Physical
Chemical
Biological

Table 6.1 expands on these headings. The connection with catchment processes appears most

Table 6.1. Elements of the freshwater habitat

Physical	Chemical	Biological
Primary Structural: bed and banks, of channel	Oxygenation	Carrying capacity, reserves and recruitment
Secondary Particulate composition, both static and transported	Nutritional status	Competition and predation
Tertiary Flow-conditions: velocity, wetted perimeter	Pollutants and their variability, concentration and loading	Introductions, managed species, fishing

commonly under the 'Chemical' heading, e.g. a river in which many expected organisms are missing, and because of oil pollution almost certainly drains a major industrial complex; a eutrophic lake which invariably, if the conditions have developed recently and relatively suddenly, drains an intensively-farmed catchment. It is perhaps more difficult to establish a connection when the cause of change in the chemical environment may have come via a physical change on the catchment, e.g faster drainage of surface waters carrying more phosphate or more acidity into rivers, for instance from upland forests. These more subtle effects are currently driving research efforts in hydrology to achieve a deeper understanding of those processes which become altered by an intervention in the hydrological cycle (see Chapter 5).

If chemical aspects of the freshwater environment have achieved prominence through the identification and management of polluted ecosystems, the physical aspects of wetlands, lakes, and rivers are more fundamental since these control the scale and heterogeneity of habitat. It is no accident that both fish biologists and geomorphologists have produced classifications for the sectors of a river long-profile; Table 6.2 compares these to draw out interrelationships. Sections 7.3 and 7.4 return to this theme.

Table 6.2. Zonation of river courses by two disciplines

Geomorphology*	Fish biology
Upstream	
Catchment: a source zone	Trout
Transfer zone	{ Grayling { Barbel
Deposition zone	Bream
Downstream	

* after Schumm (1977).

Note: The above triplet may be repeated, many times in one river basin. The main deposition zone for coarse material is often the grayling zone and for fine sediments a tidal estuary.

Physical influences tend to operate over longer time-scales of change—geomorphologists work in millennia—but there are shorter-term effects through sediment loadings to freshwater, controlled by both hydrological processes and land management. The processes of surface and channel erosion may produce periods of disturbed physical equilibrium—a turbid river where it was once clear, or a braided planform where the river once meandered.

Such periods are of marked significance for freshwater ecosystems and pose a very difficult management problem. We therefore cover the topic of restoration projects in Chapter 9.

Finally, the *biological* elements of the freshwater ecosystem are not solely the outcome of the interrelationship between physical and chemical parameters. Recruitment, migration, opportunity, and other spatially and temporally controlled biological phenomena intervene. Whilst, therefore, the physical and chemical habitat factors, particularly of stream environments, can be used to suggest what plants and animals should be present (Section 7.5) many intrinsic biological factors also have an influence. Simple food-chain relationships, including predation, are at work as are factors derived from the assemblages or network of habitats present in a river system: one polluted or destablized reach in such a system may disturb the breeding or migration of key species in an ecosystem.

River managers are also increasingly realizing that the relationship between freshwater habitats and *neighbouring wetland and terrestrial habitats* is crucial to sustained 'naturalness' in the freshwater environment. Rivers and lakes do not exist as isolated elements, divorced from their immediate surroundings because they are aqueous media; the river-bank, valley-floor, lakeshore environments all link together. The principle of *ecotones* (boundary environments) is especially relevant to banks, shores, and corridors, permitting not only biological diversity but the operation of natural processes of *modulation* (wetlands storing floodwaters) and *moderation* or purification (wetlands stripping nutrients) on runoff products of the catchment hinterland.

Finally, returning to management of the basin hydrological cycle and particularly to pollution control, it would be easy to portray the freshwater environment as deeply vulnerable to mankind's heedless exploitation, with the world's rivers and lakes heading for calamitous and permanent *degradation*. In Chapter 8 we develop the theme that vulnerability is extreme but that, seen in another way, recovery is rife: wetlands, lakes, and rivers are *resilient*. Their natural state is one of branched or zoned diversity, offering perhaps more opportunities for evasive action than other major habitats. Mankind, furthermore, has multiple opportunities to act (as in the case of the UK National Rivers Authority) as 'Guardians of the Water Environment'. The triumph of water engineering practice, formerly mainly geared to water exploitation, is increasingly focused on alternative, environ-

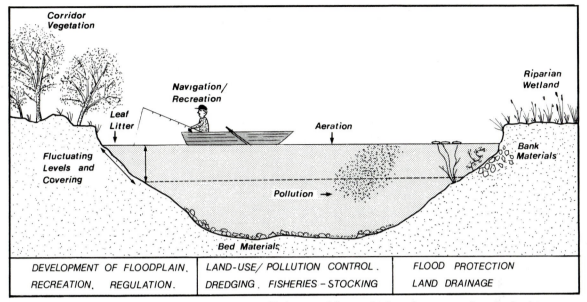

Fig. 6.1. Pressures upon freshwater systems in terms of habitat modification both in the water and on the banks.

mental aims such as flushing out pollution by reservoir releases and storing flood flows in alluvial or peat basins.

Fig. 6.1 illustrates the three major components of the freshwater habitat we have developed in this link section and indicates the management relationships to be addressed, preferably on a whole basin scale.

7. River and wetland systems

This text has advocated a holistic approach to river basins and their component hydrological processes which help control freshwater environments. A century and a half ago Victorian scientists were addressing 'big questions' in natural science; these often turned out to be of childlike simplicity: 'Why are creatures different?'; 'How do glaciers flow?' In the intervening years we have miniaturized the components of an answer whilst, simultaneously, distancing ourselves from the questions; this has been the necessity of truly scientific (experimental) enquiry. The unifying reassembly pattern can now be offered by natural systems (including ecosystems) arguments and by the power of computing (see Chapter 9). We do, however, need those simple questions to be posed and re-posed.

7.1 What is a river? A lake? A wetland?

As runoff gathers by the coalescence of surface rills, or subsurface fissures under the force of gravity, it has an increasingly visible presence in the landscape (groundwater as springs or spring-lines), and brings about morphological change. The geomorphological essence of environmental change means, however, that not all the elements, even of a water-worn, fluvial landscape are in equilibrium with present conditions. Thus we observe *abandoned river channels* and lake basins, the victims of climate change; we also have landscapes where the hydrological regime has yet to profoundly modify the topographic legacies of past conditions (e.g. zones of recent glaciation).

Whatever the state of *landscape equilibrium* and change there are simple hydrological descriptors of the three key components of wet habitat: scale, variability of water level/content, and flow-through time; all are interrelated. Table 7.1 develops these variables.

The essence of rivers, lakes, and wetlands is, however, their combination and connectivity in forming wet habitat (Fig. 7.1(a)). For management purposes it is essential to see them as component

Table 7.1. Key components of wet habitat

Habitat	Scale/ Geometry	Water level/ Content	Flow-through time
River channels	Continental/ linear	Variable, mainly water-filled	Minutes to weeks
Ponds/lakes	Local to sub-continental/ areal	Fluctuating, water-filled	Days to years
Wetlands	Sub-basins, coastal. Often linear or concentric	Fluctuating, saturated	Minutes (surface) to years (subsurface)

parts of the *extensive* natural resources of river basins; disturbing the connectivity, e.g. between stream lengths (by damming), or between streams and wetlands (by flood protection), can threaten the performance of the system as habitat. This connectivity and interdependence makes it doubly unfortunate that the knowledge base for river-basin management tends to be derived from single-discipline research active at only one scale (see Fig. 7.1(b)).

7.2 Quantifying the extent of, and damage to rivers, lakes, and wetlands

Whilst most nation states catalogue and calibrate the extent of mineral and energy resources for example, their approach to wetness has been extremely variable. In those cases where expertise and resources have been available to make *hydrometric measurements* these have mainly been in support of the exploitation of *water resources* rather than of pollution control or conservation. Conventional hydrometry often fails to measure important components of habitat such as stream-bed velocity and has tended to ignore many lakes and most wetlands. We can derive global estimates in hydrology only by extrapolation across national boundaries or by global remote sensing but the

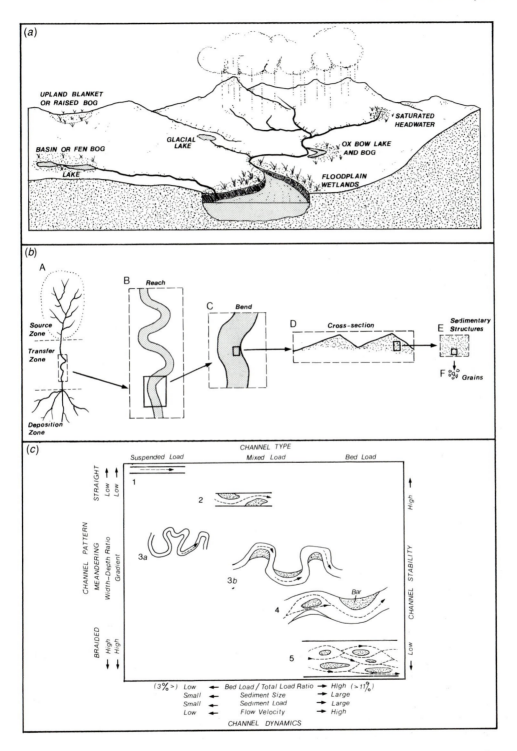

Fig. 7.1. Systematic treatment of the river system: (a) pictorially; (b) as a series of study scales in geomorphology/geology; and (c) as a morphological classification (after Schumm, 1985).

nation state is an extremely important unit of environmental management and policy (Johnston, 1989).

Chapter 1 logs the data acquired by taking global hydrometric approaches to the hydrological cycle. This section extends that approach and provides examples of national assessments of the broader wet environment.

Wetlands are among the world's most productive ecosystems but are also highly sensitive to exploitation; Maltby (1986) uses the title *Waterlogged Wealth* for his world survey of the remaining assets and their conservation. He catalogues the innate value of waterlogged habitat.

Some wetlands can produce up to eight times as much plant matter as an average wheat field; wetlands sift dissolved and suspended materials from floodwaters . . . maintaining water quality; wetlands protect coasts and inland areas from floods. Wetlands are the last truly wild and untouched places on earth. (p.9)

The natural products of wetlands include fuel, construction materials, food, drugs, paper, and oils. Peatlands alone cover 5 million km^2 and extend to all continents; Canada and the Soviet Union have over 1 million km^2 each! Moore and Bellamy (1974) provide a concise guide to the incredible variety of physiographic manifestation and function of the peatlands.

If nation states have been slow to provide environmentally based resource assessments of freshwater habitats they have been even more backward, because of the complexity, and hence expense, of the task, especially to assess *damage* and *loss*. Paradoxically, it is the simple direct observation of the naturalist which often provides the best (and earliest) indication of a stressed environment; the problem becomes one of taking action on the basis of 'merely' qualitative evidence. The technological orientation of developed and developing states normally demands that decisions are based upon 'sound', 'hard' numerical data, the problem being that these are extremely difficult to gather. A fully equipped monitoring station to continuously monitor river-water quality and to count migratory fish is a very expensive undertaking. Fortunately, however, as this and the next chapter reveal, assessments based upon codified and repeatable *biological surveys* are now available, as an increasingly sophisticated compromise between chemist and naturalist.

The capacity to create an inventory of the freshwater resource for an area of the Earth's surface is highly dependent on how well it is mapped or remotely sensed. For great Britain, geomorphologists have employed the high-quality map base provided by the Ordnance Survey to calculate, for example, stream network indices such as *drainage density* or the simpler *stream frequency* (number of stream-channel junctions per unit catchment area).

Smith and Lyle (1978) took this approach a stage further in an attempt to quantify the number as well as the distribution of stream channels and lakes in Britain. Their results are shown in Table 7.2. The basis of their calculations for rivers is the *stream-ordering* system in which a single, unbranched headwater channel is labelled first order with the order of subsequent downstream links being incremented each time two streams of the same order join. The ordering system is a convenient way to organize information about river networks, conveying considerable positional information; with the addition of further codes exact locations in the network can be specified and become the basis for hydrological river pollution or habitat databases.

Table 7.2. The systems resource: the estimated total number of lakes and streams in different size groups in Great Britain

Lakes		Streams	
Area (km^2)	Number	Order	Number
Mainly <0.004	56 176	1	146 853
Mainly 0.004–0.25	4 416	2	36 534
0.25–1	848	3	8894
1–4	177	4	1937
4–16	51	5	386
6–32	9	6	66
>32	4	7	4

Source: Smith and Lyle (1978).

Clearly stream order can act as a very simple surrogate for those elements of the river habitat which vary within the stream system; however there are both quantitative and qualitative elements which must be added to any more refined system of channel classification which could be used to, first, predict what biota should be present and, secondly, indicate damage or stress by comparison of predictions with field-survey data of what is present (see Section 7.5.2).

Newson and Harrison (1978) took *channel classification* a stage further in purely descriptive terms, both mapping and tabulating the major categories of channel within the confines of a small upland catchment in Wales (see Table 3.4). Whilst broadly applicable to the conditions of upland areas in the British Isles this approach is, however, not generalizable to all fluvial landscapes. Once again it is geomorphologists who have attempted, in terms of planform geometry of channels to ask the simple question, 'What is a channel?' (see Section 3.6.1). The most widely accepted answer is that provided by Schumm (1977). This combines visually (see Fig. 7.1(c)) the elements of:

sediment load
stream power
cross-section geometry
stability

to produce a working classification.

In quantifying *drainage* to the water environment one may take either a relative or absolute standpoint. Once again, progress by various nations depends essentially on their ability to quantify the 'stock' of rivers, lakes, and wetlands, monitor for change in the stock, and classify broadly in order to summarize the situation. In the UK, for example, the work of Smith and Lyle (1978) concluded that fresh waters covered a mere 1.04 per cent of the land surface; activity on the remainder of that surface can control, as we have seen in Part I, the hydrological performance of freshwater systems and their habitat status. It is interesting, therefore, that the UK has had, for nearly twenty years, a system of *classifying water quality* from 'good' to 'very poor' in its river networks (see Chapter 8). Progress in combating pollution is measured by assessing the changing lengths, and proportions of the total network, in each class. Lakes are not given comprehensive surveys of a similar nature but those in the uplands of Britain, in Scandinavia, and Canada have come under recent scrutiny as a result of the threat of *acidification* (Mason 1990; Stevenson *et al.*, 1991); in recent years, too, the threat of *algal blooms* on lowland open-water bodies has led to regular surveys (National Rivers Authority, 1990).

The effects of direct intervention to manage river channels to improve *flood capacity*, or their flows, as part of reservoir regulation have been assessed by geographers in the UK. Brookes *et al.* (1983) mapped the distribution of some 8500 km of channels on which comprehensive engineering schemes had been carried out and a further 27 000 km which were maintained by regular excavations of silt or removal of vegetation (see Brookes, 1988).

In reviewing the impact of river regulation in Britain (see also Section 5.8.4) Petts (1988) calculates a regulated catchment area of almost 90 000 km^2 in the UK—around 40 per cent of the land area. Between 10 and 40 per cent of gauging-station records in the UK (793 individual sites) are said to be affected by artificial flow manipulations. Volume 2 of the journal in which Petts writes contains a thorough review of the impacts of regulation on flow regime, sediment transport, fisheries, recreation and conservation (see also Petts, 1984; and Section 5.8 here).

The international dimensions of the campaign to conserve wetlands has yielded a much more comprehensive database on damage to wetland ecosystems; in this case damage tends to be absolute and often irreversible whereas rivers and lakes can frequently be 'cleaned up' or restored. The UK Nature Conservancy Council charted the damage to 95 per cent of lowland raised mires, 50 per cent of fens, valley and

---

Box 7.1. *Major causes of wetland loss and degradation*

HUMAN THREATS

Direct

Drainage for crop production, timber production, and mosquito control.

Dredging and stream channelization for navigation channels, flood protection, coastal housing developments, and reservoir maintenance.

Filling for dredged spoil and other solid waste disposal, roads and highways, and commercial, residential, and industrial development.

Construction of dikes, dams, levees, and seawalls for flood control, water supply, irrigation, and storm protection.

Discharges of materials (e.g. pesticides, herbicides, other pollutants, nutrient loading from domestic sewage and agricultural runoff, and sediments from dredging and filling, agricultural and other land development) into waters and wetlands.

Mining of wetland soils for peat, coal, sand, gravel, phosphate, and other materials.

Indirect

Sediment diversion by dams, deep channels, and other structures.

Hydrological alterations by canals, spoil banks, roads, and other structures.

Subsidence due to extraction of groundwater, oil, gas, sulphur, and other minerals.

NATURAL THREATS

Subsidence (including natural rise of sea-level), droughts, hurricanes and other storms, erosion, and biotic effects.

Source: Maltby (1986).

Fig. 7.2. Loss of Wetlands in Britain: (a) the Fens of East Anglia; (b) upland blanket bog (after Nature Conservancy Council, 1984).

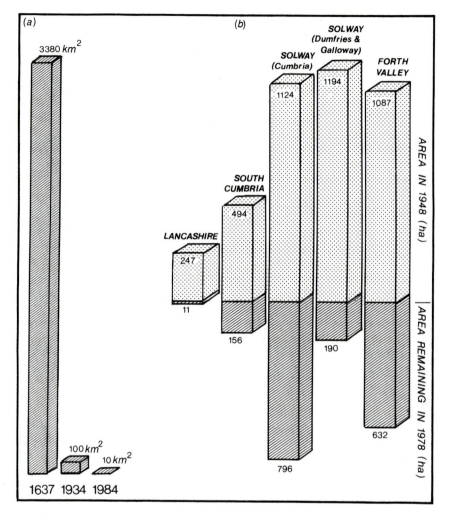

basin mires and, between 1637 and 1984 a reduction in area of the East Anglian Fens from 3380 km^2 to 10 km^2 (see Fig. 7.2). The Fens are also 'wasting' vertically; Richardson and Smith (1977) calculate rates of drainage-induced peat shrinkage at the famous Holme Fen Post, installed level with the ground in 1852 and now (1971) exposed by 0.38 m. Elsewhere in the Fens the authors measure current shrinkage rates of between 0.27 and 3.09 cm per year. Baldock (1984), in a survey of the damage threats to 230 wetlands in France, the Netherlands, Ireland, and the UK concludes that the era of large-scale land reclamation for agriculture is largely over but that smaller scale operations are 'nibbling away at the edges of poorly protected sites' (p. 158). As a result it is Baldock's view that, 'if the decline of wetlands continues on the present scale, we can expect

some very substantial losses and a serious impoverishment of the European environment within a few decades' (p. 160).

Clearly we are moving into a very dynamic era of changing attitudes and climates for wetlands; the agricultural threat has abated in Europe because of the redirection of funding for 'improved' production; however, land development for industry, especially at the coast, continues and flood defence in response to 'greenhouse effect' hazards may also jeopardize the wet habitat. Meanwhile, as Maltby illustrates (Tables 7.3(a) and (b), and Box 7.1), in the United States the rate at which wetlands have beeen lost continues to be spectacular and there is not likely to be any relief from the major human threats to the remaining sites.

Table 7.3(a). Wetland losses in selected US states and regions

State/region	Original wetlands (000 ha)	Remaining wetlands (000 ha)	% loss	Date
Iowa—natural marshes	930	10.7	99	1981
California	2 000	182	91	1977
Nebraska—rainwater basin	38	3.4	91	1982
Mississippi alluvial plain	9 700	2 100	78	1979
Michigan	4 500	1 300	71	1982
North Dakota	2 000	810	60	1983
Minnesota	7 450	3 500	53	1981
Louisiana—forested wetlands	4 570	2 300	50	1980
Connecticut—coastal marshes	12	6	50	1982
North Carolina—pocosins	1 000	610	40	1981
South Dakota	810	525	35	1983
Wisconsin	4 000	2 700	32	1976

Source: Maltby (1986).

Table 7.3(b). Recent rates of wetland losses in the conterminous United States, approx. 1955–1975

State/region	Loss rate (ha/year)	Date
Lower Mississippi alluvial plain	66 000	1979
Louisiana—forested wetlands	35 300	1980
North Carolina—pocosins	17 600	1981
Prairie pothole region	13 400	1969
Louisiana—coastal marshes	10 000	1982
Great Lakes Basin	8 100	1981
Wisconsin	8 100	1976
Michigan	2 600	1981
Kentucky	1 450	1983
New Jersey—coastal marshes	1 250	1973
Palm Beach County, Florida	1 240	1982
Maryland—coastal wetlands	405	1983
New York—estuarine marshes	300	1972
Delaware—coastal marshes	180	1983

Note: Examples of recent wetland loss rates in the USA: agriculture 87%; urban development 8%; other development 5%.

Source: Maltby (1986).

7.3 Physical, chemical and biological elements of habitat

Smith and Lyle (1978) devised a diagrammatic cube by which to classify fresh waters in Britain on the basis of size, altitude, and geology (modified here in Fig. 7.3). It is of broad guidance only, usable in the way suggested for the matrix of wetland types offered by Newson (1992), i.e. as a basis for impact assessment (see Section 7.7 and Box 7.5).

The important classificatory problem in assessing the habitat status of rivers and lakes has been to provide a zonation of the linear continuum of the river system (Vannote *et al.*, 1980) or the depth of lakes and reservoirs.

7.3.1 Downstream-zoning of rivers

We have already referred to a degree of convergence between the approaches to rivers by geomorphologists and biologists (Chapter 6). A further

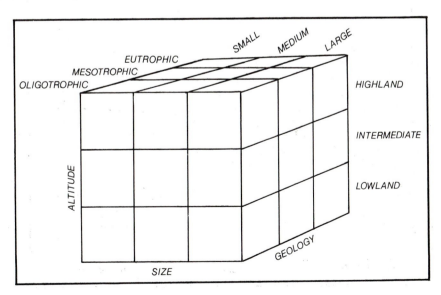

Fig. 7.3. The 'habitat' cube for freshwaters (after Smith and Lyle, 1974).

similarity is revealed by the fact that both groups of researchers have now moved from purely zonal treatments, in which the precise local energy conditions were largely irrelevant, to assessments based upon building blocks formed from detailed studies at-a-site. In the case of biology this has meant a move towards the *trophic status* of waters within river reaches and to assessment of the *conservation value* of assemblages of organisms—their abundance and diversity. However, for geographers and other Earth and environmental scientists approaching hydrobiology for the first time, the traditional zonations are worthy of a brief treatment.

The principal zonations were based upon *fisheries* and bore the names of the characteristic fishes of each zone. Nevertheless a generation of biologists researched the changes in controlling physical, chemical, and biological parameters throughout the longest streams in river systems, producing curves such as those described by Ullman (1979) (see Fig. 7.4(a)).

Use of hydrological data has on occasion refuted some of the assumptions of the zonation model. For example Ledger (1981), using current-meter data for the River Tweed in Scotland, showed that *mean* velocity increased downstream under conditions other than flood but decreased slightly at mean annual flood and above. Even this contribution is based upon *mean* velocity whereas it is clearly the

bed velocity and turbulent components of the mean downstream velocity which most impact on the local habitat of plants and animals.

Jeffries and Mills (1990), commenting on the traditional zone model suggest that its most significant breakpoint appears to relate to water temperature and to oxygen demand. Above the lowland-reach temperatures seldom exceed 20°C and there are few 'oxygen sags' (see Section 8.3.2). Below this breakpoint biota are adjusted to higher temperatures and a shortage of oxygen from time to time. These authors eventually compose a ninefold classification based upon location in the stream continuum and trophic status (Table 7.4).

The alternative concept of the *river continuum* is developed by Vannote *et al.* (1980). It emphasizes the components of river habitat largely on the basis of trophic status, the many dimensions of which may fluctuate in a quasi-zonal way, producing significant combinations for the biota in any reach (Fig. 7.4(b)). Further specifications of habitat features can be presented in terms of biological needs at various stages of the life cycle or at different times of the year, e.g. Table 7.5.

There is clearly no single, correct way to classify river habitats throughout the system; the increasing utility of biotic assessments for guiding conservation, pollution control, and fisheries management means that there are bound to be 'horses for

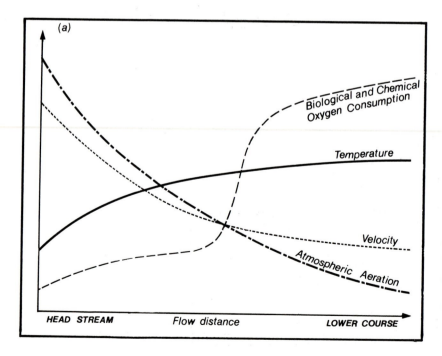

Fig. 7.4. The stream environment and habitat: (a) as varying downstream (after Ullman, 1979); (b) *opposite* as an ecosystem in cross-section (after Boon, 1991).

courses'. Each indicator of biotic abundance or diversity pools a large measure of habitat information and there are now obvious signs that this information includes the contribution to channel habitats made by neighbouring terrestrial habitats in the *river corridor*.

7.3.2 Downward-zoning of lakes

Lakes are not totally immobile bodies of water—they have inflows, outflows, and are disturbed and mixed by waves and currents. However, they have much greater depth in most cases than rivers. This depth has a number of important physical, chemical, and biological effects (for a general introduction to lakes see Burgis and Morris, 1987). At depth change is very slow and a physico-chemical zonation develops which is termed *stratification*.

A very obvious physical effect with depth in any water body is the scattering and absorption of *light*. The *euphotic zone*, which varies between about 0.5 m and 20 m, is that bounded at its base by

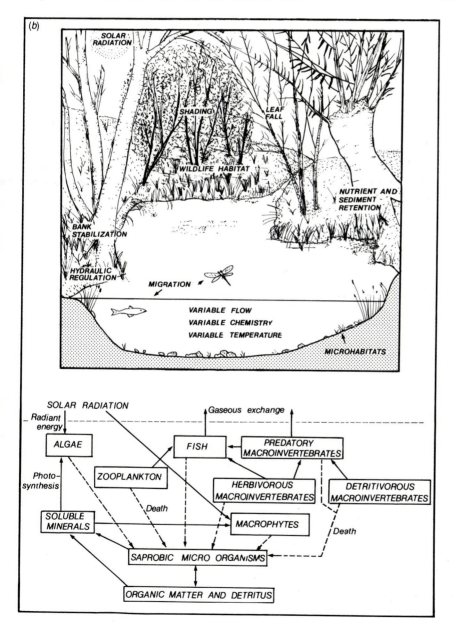

Table 7.4. General vegetation patterns along oligotrophic, mesotrophic, and eutrophic river systems

River section	Trophic status		
	Oligotrophic	Mesotrophic	Eutrophic
Head-streams	Upland, resistant geology, spatey, coarse substrate.	Such rivers typically start from spatey upland streams and have a similar fauna to oligotrophic systems. Acidic flora generally absent and patches of herbs and grasses thrive in protected refuges.	Generally lowland rivers on soft geology or alluvium. The headstreams are often shallow gradient and nowadays often ditched and diked in areas of intensive agriculture. Many herbs and grasses flourish and can clog the streams, e.g. watercress, water-celery, reed grass, reeds, bur-reeds, starwort, water-mint.
	Few vascular macrophytes in channel. Mosses and algae dominate. Fringing plants include herbs and grasses, e.g. *Agrostis stolonifera* and reed grass. Common in uplands: acid bog species, e.g. bog pond-weed, sedges, and cottongrass.		
Middle reaches	Channel macrophytes remain sparse. Fringing herbs may be abundant, washed out and recovering each year, e.g. brooklime, water-celery, watercress. Mosses and algae still frequent.		As channel size increases the herbs are replaced by genuine submerged channel species with fringing grasses, reeds, and emergents typical, e.g. milfoils, floating bur-reeds, lilies, arrowhead, water-plantain, reeds, reed-grass, starworts. Many such rivers are now polluted with excessive plant nutrients and are turbid. In such cases channel species may be limited and algal growth abundant.
		Softer rocks, lower gradient, deposition and richer nutrient supply all allow good vascular flora to develop. Many herbs, grasses and channel species, e.g. crowfoots, broad-leaved and curly-leaved pond-weeds, floating and erect bur-reed, brooklime, forget- me-not, watercress.	
Lower reaches	Lower gradient and deposition allow vascular macrophytes to survive in channel crowfoots and fringing herbs, and grasses abundant. Mosses and algae may now be out-competed.		

Source: Jeffries and Mills (1990).

receipt of 99 per cent of incoming, surface light. Below this plants cannot photosynthesize and animals dependent on them must remain in or visit the euphotic zone to feed, or await their death and sinking.

Water bodies are slow to heat and cool; conduction of surface heat to depth is slow in the absence of heating. *Thermal stratification* therefore occurs seasonally in many lakes, the stratification relating also to temperature-induced density variations. Water is most dense at a temperature of 3.94°C; thus in summer deep lakes stratify as shown in Fig. 7.5(a). In autumn as surface water cools, and bottom waters have at last warmed, a *turnover* occurs which can have profound water-quality effects: bottom waters have become *deoxygenated*, may be turbid from sediment inflows, and may contain *dissolved metals* such as iron and manganese. Whilst much of the temperate zone has two turnover peri-

ods per year, with ice forming at the surface in winter (Fig. 7.5(b)) those with mild oceanic winter conditions experience only the autumn turnover. Burgis and Morris (1987) put the minimum depth for turnover behaviour as 10–15 m: shallower lakes do not stratify.

The zones of a stratified lake are known as *epilimnion*, *thermocline*, and *hypolimnion* (Fig. 7.5(a)). There are, however, other characteristics of lake depth at *lake margins*—the littoral zone—which have an important bearing on the biological interest of lakes and also of larger rivers. The combination of depth, shore profile, and water-surface fluctuations markedly affects the development of habitat through growth and decline of vegetation as a *hydrosere* (Fig. 7.6(a)) and through modifying the feeding habits, for example, of water birds (Fig. 7.6(b)).

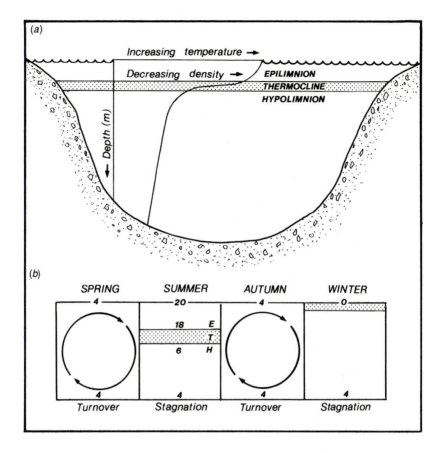

Fig. 7.5. The stratification of lakes in (a) cross-section, and (b) over the seasons (after Ullman, 1979). Units are degrees Celsius.

Table 7.5. Principal habitat features important to fish

Biological process	Governing habitat features
REPRODUCTION	
access to spawning areas	Provision of suitable depths and water velocity, absence of barriers to movements
spawning	Suitable spawning substrate
incubation of eggs	Stability of substrate, provision of adequate temperature, oxygen supply and water movement
FEEDING AND GROWTH	
availability of food organisms	Bankside and aquatic vegetation, substratum suitable for invertebrate production, supply of allochthonous organic material
best use of energy for maintaining position and food gathering	Cover and shade, e.g. rocks, tree trunks, diversity of flow type, riffle-pool sequences (in trout streams), aquatic and bankside vegetation, appropriate temperature range
SELF-PROTECTION	
from physical displacement by current from predation from competition with their own and other species	Shelter and visual isolation, e.g. varied bed-profile undercut banks, rocks, tree trunks, roots, accumulated debris, aquatic vegetation and, for fry and juveniles, weedy shallow marginal slacks, backwaters

Source: Milner (1984).

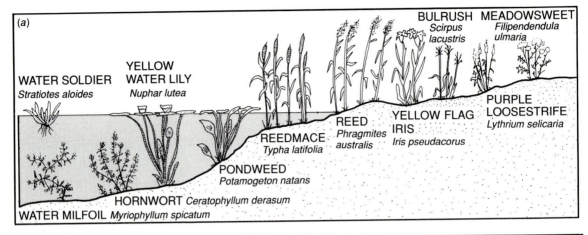

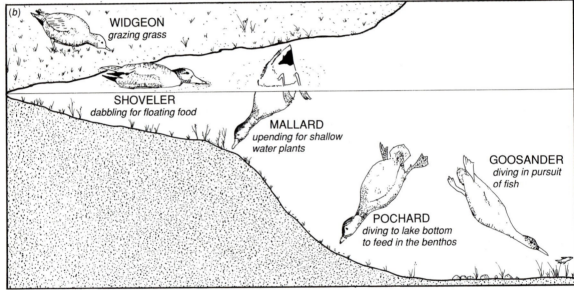

Fig. 7.6. Marginal freshwater habitats of shorelines. (a) Plant zonation (after Burgis and Morris, 1987). (b) Importance of variety for diversity of duck feeding habits (after Burgis and Morris, 1987).

7.4 Individual habitat elements and their effects on biota

Brian Moss in his book *Ecology of fresh waters* has a chapter entitled 'On living in water' (Moss, 1988); the purpose of this section is to provide a brief introduction to the conditions of habitat in rivers and lakes (the conditions for wetlands are described in Section 7.7).

Two immediate problems of life in water are the *maintenance of ionic balance* (for bodies and parts of bodies which are permeable) between cell solutions and the often fluctuating chemical content of the water outside, and the winning of an adequate oxygen supply from a medium in which it is in relatively short supply. Both have a bearing on pollution of water and are developed further in Chapter 8.

Whilst still-water or *lotic* habitats are characterized by temperature, light, and chemical concentrations (particularly at turnover) the flowing-water or *lentic* habitat is characterized by a much greater, more variable interaction between water and substrate. The physical substrate of a stream channel consists of the mineral material of bed and banks, together with any plant growth which may have colonized it.

There is clearly an interaction between substrate, gradient, and flow volume in various parts of a river system. The outcome of this interaction is a characteristic velocity and turbulence distribution controlling the need and opportunity for *anchorage* for plants and animals, the distribution of *oxygen* and *food* supply in the reach, and the location of *shelter* and *protection*.

The so-called *Instream Flow Incremental Methodology* developed by the United States Fish and Wildlife Service (Gore and Nestler, 1988) seeks to elucidate by field-data collection the hydraulic parameters which underlie habitat-suitability for target species before those conditions become altered, e.g. by river regulation. Since it is impossible to collect data to give a total picture of the river system, fieldwork is based on calibrating conditions in the cells of a hydraulic model which will generalize the gross changes in habitat availability for each species. Gore *et al.* (1989) extend the technique to the dynamic changes of habitat for a fish called the *banded sculpin* during the passage of a wave of increased flow during hydroelectric-power generation on a regulated river in Tennessee. The weighted usable area for the sculpin decreased during the artificial flood wave and the operation of the turbines can be constrained, if necessary, to reduce the impact of this stress.

7.5 Biotic indicators of 'naturalness'

In Section 7.3 we highlighted the fact that biologists had moved from the rather static classification of river zones to schemes of classification of habitats which had a predictive, and hence applied, value. Such procedures use field surveys of plant or animal distributions under 'natural' conditions to relate to habitat factors—these plants or animals are then said to be *indicators* of the operation of particular habitats.

Clearly there are two major candidates for such systems: plants and animals, and two major ways of using them—as individual indicators or grouped statistically.

7.5.1 Plant indicators in rivers

Jefferies and Mills (1990) wax lyrical on the diversity of freshwater plants:

The verdant plants that spill over many waterways, the tangled trails of river weeds, the whispering reed beds or malignant slime that surface on a garden pond all conjure up images of freshwaters focused around the plant life. (p. 63)

For non-biologists interested in plant indicators in freshwaters there are two prominent divisions of this myriad of plant life: *macrophytes* (visible to the naked eye) and *phytoplankton* (microscopic). The former, including the larger algae, mosses, liverworts, and flowing plants are the basis of two prominent approaches to river habitat classification whilst the latter are increasingly used as diagnostic of artificial changes in the trophic status (especially of lakes).

Haslam (1978) takes a very simple approach to lentic macrophytes: 'The vegetation types of watercourses may be classified according to the flow pattern of their channels and the geology of their catchment area' (p. 7). She uses the interrelationships between geology, physiography, and climate, taken for granted by geomorphologists and hydrologists, as a framework from which broad plant groupings can be formed. For example, having determined that river vegetation is affected by the 'force of the water flow and its variations' she lists the determining factors of this flow:

- height of the land, of the hills in general;
- height of the hills above the stream, i.e. the fall from hill to channel;
- downstream slope of the channel;
- seasonal distribution of rainfall;
- total annual rainfall;
- porosity of land surface;
- distribution of springs.

Clearly this is a qualitative restatement of some of the correlations between catchment variables uncovered by flood prediction research (see Table 4.5); such an approach works in Britain where there is a broad natural correlation between rock hardness, climate, and relief.

Furthermore Haslam's approach to the plants themselves is qualitative, there being 'only about fifty river plants which are both widespread and of diagnostic importance in Britain'; they can be placed fairly easily into five groups. The first group is the most important since it arranges river plants according to their nutrient status. The second group classifies the important water crowfoots (*Ranunculus spp.*), the third fringing herbs, the fourth 'left-overs' and the fifth bank-plants generally absent from the channel.

Histograms of species abundance in relation to broad habitat categories (see Fig. 7.7(a)) of flow, substrate, and river geometries are used to confirm the utility of species groupings as indicators; nutrient conditions are depicted as star diagrams of water analyses for typical sites (Fig. 7.7(b)). The

Fig. 7.7. Haslam's presentation of the relationship between river habitat and characteristic species; (a) as histograms of abundance in relation to flow regime; (b) as 'star diagrams' in relation to chemical water quality (after Haslam, 1978).

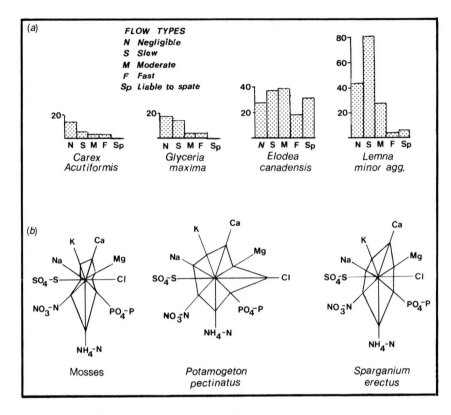

approach is extended to North American streams. However, there is no repeatable predictive power in this technique; its biggest service was to raise in the minds of river managers the impact of their actions in altering certain important aspects of an important environment.

Later work by Haslam and Wolseley (1981) retained the generally descriptive, amateur survey approach but provided a *predictive tool* in the form of a dial upon which the naturalist moves pointers for geology, physiography, and cross-sectional geometry into a position where the typical plant groups for the river in question are indicated. An assessment of damage is made by comparing the plants present with those indicated as likely; a *pollution index* is derived based upon the difference between observed and predicted plant species in reaches unaffected by physical damage. It is interesting to note that Haslam and Wolseley list nineteen sources of damage other than pollution:

substantial shade at sides or over whole channel;
visitor trampling, paddling or swimming;
cattle disturbance, trampling or grazing;
boats;
recent dredging;

recent weed cutting;
herbicides sprayed on emerged (or floating) species;
aquatic herbicides used in the water;
roadworks affecting channel or temporarily causing extra mud etc. to wash in;
bed made of concrete, boulders or other coarse substrates;
bed made of man-made unstable substrate;
undue turbulence or deep water caused by bridge piers or other structures;
unduly steep banks for the type of channel;
unduly shallow, or wide, or deep;
lowered water level in previous year (due to excavation);
summer drying;
fierce spates;
very swift normal flow;
regulation of flow (through reservoirs, water transfers, gates, etc.) leading to more irregular discharges.

A further purpose for the work by Haslam is, in the case of both books, the education of those managing river vegetation *directly*. The growth of macrophyte vegetation in channels has both advantages

Table 7.6. Channel vegetation: the balance sheet

Advantages	Disadvantages
1. Pleasing appearance	1. Restriction of flood capacity
2. Aeration and purification	2. Interference with angling
3. Stabilization	3. Difficulties of navigation
	4. Obstruction of structures by detached vegetation

and disadvantages (apart from the purely ecological). Table 7.6 summarizes these. In lowland rivers of high nutrient status the growth of macrophytes will need clearing to prevent flood risk growing at the same rate as the 'weeds' obstruct the channel. Careful use of chemicals and sensitive deployment of specially modified boats are techniques employed to prevent the creation of a sterile channel choked with dying vegetation.

A major qualitative breakthrough in the assessing of river vegetation types for conservation purposes came after a five-year survey of 1 km river reaches by Holmes (1983). He entered the species lists for 1055 sites in the TWINSPAN computer model which identifies significant groupings; 56 groups were identified which Holmes reduced to 4 major classes based largely on geology:

A Lowland and rich geololgy;
B Sandstone, millstone, and hard limestone;
C Resistant geology;
D Upland acid and nutrient poor.

Having recorded the plants of 150 rivers in Britain it was possible to map the major classes (Fig. 7.8).

7.5.2 Invertebrate indicators

In an extremely comprehensive survey of the utility of biological indicators Hellawell (1986) has pointed up the popularity of the macroinvertebrates (Fig. 7.9(a)).

From the earliest days of modern concern over polluted streams an educational use has been made of those indicator species which children might find whilst fishing in their local stream; Fig. 7.9(b)) shows how seven species of freshwater invertebrate animal may be used in an extremely tentative but practical classification of water quality (Advisory Centre for Education, 1971).

Only recently, however, has the predictive but qualitative approach taken by Haslam to macrophytes and the comprehensive and stastistical approach of Holmes to the same organisms been combined in a purposeful geographical and statistical survey of British stream macroinvertebrates.

For many years the number and diversity of stream invertebrates has been used in purely biological studies or, more recently, to support pollution surveys (see Hellawell, 1986; Metcalfe, 1989; British Ecological Society, 1990) (see also Plate 7.1); indices and scores (e.g. Trent Biotic Index, Chandler Biotic Score, Biological Monitoring Working Party Score) have been used to summarize the ecological water quality of river reaches. However, whilst sample collection may be relatively easy (commonly the river bed is disturbed by the operator *kick sampling* the bed upstream of a plankton net) identification of invertebrates is very difficult. There have been promises of automation via image-processing techniques coupled to microscopy but the fact remains that amateur natural historians, the target operators of Haslam's plant schemes, cannot perform detailed analyses of invertebrates.

Fig. 7.9(c) illustrates the procedures adopted by a computer-based predictive system devised by the UK Institute of Freshwater Ecology. Known as RIVPACS, it assembles the results of surveys of 268 undisturbed river sites, relates the statistically established classification of invertebrates to the controlling habitat measurements and offers predictive equations based upon them (see Box 7.2). Reviewing RIVPACS (River Invertebrate Prediction and Classification System, described in detail by Wright *et al.*, 1989) the British Ecological Society (1990) is realistic. Whilst deviations between observed and predicted taxa can be used as a measure of impact, and the presence and absence of certain species used as a clue to the precise habitat change, use of RIVPACS requires the addition of much specialist judgement at the borders of major habitats or where habitats are in transition through natural environmental processes.

The advantages of using macroinvertebrates to assess 'naturalness' are said by BES to be:

- macroinvertebrates include a wide range of species providing a good *range of sensitivities*;
- they are relatively long-living and immobile, therefore offering an *integrated record* of episodic impacts such as pollution events;
- if classification is held at the level of taxa, *identification* is relatively easy and *rapid*.

7.5.3 Fisheries

Freshwater fishing has a very long history in Britain, first as a quasi-agricultural activity using traps and weirs but latterly as a pastime with rod and line (see Burton, 1982). One might claim that the use of an 'angle' or hook operated to conserve fish stocks from the time of the first celebrated treatise on fishing with a rod (Walton, 1653). However in the three centuries since Walton wrote, river stocks have declined through pollution and angling has grown to become our most popular participatory sport. The corollary of these developments is that anglers have sought to establish both their individual success and the magnitude and 'health' of the basic resource—fish stocks—by making *records of catches*. In addition specialist survey techniques such as *electro-fishing* (Plate 7.2) have grown up to allow sampling of the status and trend of fisheries.

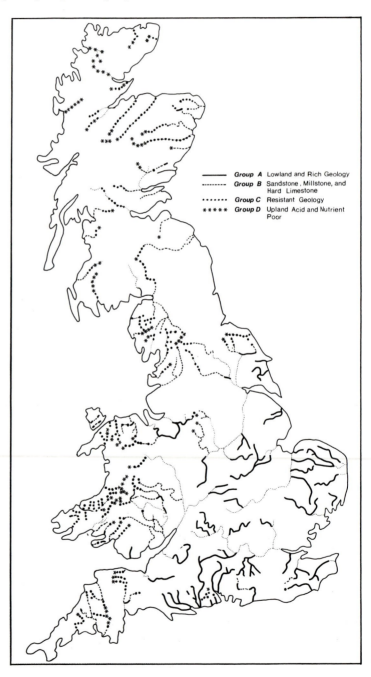

Fig. 7.8. Holmes's classification of British rivers based on statistical analysis of macrophyte plant communities.

Fig. 7.9. Use of stream biota as pollution indicators. (a) The popularity of various taxonomic groups (after Hellawell, 1986). (b) The use of stream invertebrates to delimit four river classes (A–D in declining quality) (after Advisory Centre for Education, 1971). Species are: A. Stonefly nymph, B. Freshwater shrimp, C. Chironimid larva, D. Sludge worm. (c) The use of a formal invertebrate classification in 'RIVPACS' (after British Ecological Society, 1990).

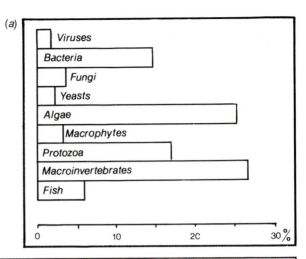

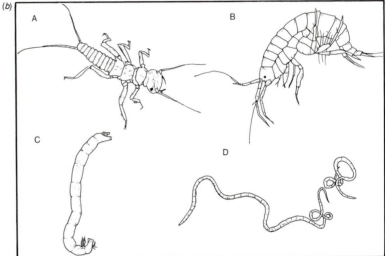

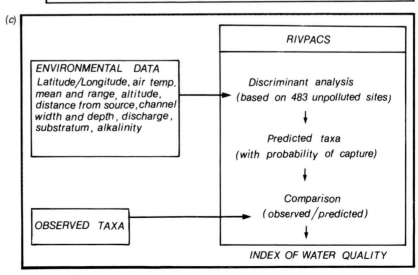

Plate 7.1. Inspecting the products of 'kick sampling' for invertebrate surveys.

The quantification of fish stocks is, however, a relatively blunt tool in assessing 'naturalness' and deviations from it in terms of freshwater habitat. Fish numbers may be affected by a number of indirect and artificial factors, principal of which are:

- the growing importance of *stocking* rivers and lakes with hand-reared young;
- factors affecting the *recruitment* of migratory fish to a particular river or reach (including coastal fishing activity).

The use of fishery records and fish surveys can, therefore, be imprecise and open to conjecture. Nevertheless, a number of recent uses of fish numbers as environmental indicators reveal their potential. For example, much of the current management of air pollution to reduce the acidification of surface waters was stimulated by the decline or absence of fish from lakes in Scandinavia, Canada and parts of Britain. Fig. 7.10(a) illustrates the careful use of control data to indicate a reducing fish population in southern Norway.

Fishery records have also been used to indicate some of the effects of river regulation on freshwater habitats. Mann (1988) stresses a balance between deleterious effects and compensating benefits (such as fish-pass structures, improved drought flows, and

hatcheries). However, Williams (1989) links a decline of salmon in the Columbia River, northwest USA, to the design and operation of hydroelectric dams.

Recent work on the River Tyne in northern England has also pointed to the effects of hydroelectric-power operations on salmon stocks in the reaches immediately downstream of Kielder Dam; the causal links to the decline of juvenile fish appear to be stream temperature and sedimentation changes. Lower down the Tyne system fish numbers have considerably recovered following improvements to sewage treatment (Fig. 7.10(b)).

7.6 Stressed river and lake environments

We have referred on a number of occasions in this chapter to pollution, drainage, and the vulnerability of freshwater ecosystems; Haslam's list of nineteen forms of non-pollution damage is indicative of this

---

Box 7.2. *An example of the use of RIVPACS predictions at a specific site*

River: MANNINGTON BROOK Site: NEWMAN'S LANE

(a) *Environmental data used:*

Water width (m)	2.40
Mean depth (cm)	20.00
Substratum composition	
Boulders + cobbles (%)	0.00
Pebbles + gravel (%)	56.70
Sand (%)	20.70
Silt + clay (%)	22.70
Mean substratum (phi)	0.39
Altitude (m)	20.00
Distance from source (km)	7.90
Slope (m km^{-1})	2.60
Air temperature range (°C)	12.32
Mean air temperature (°C)	10.60
Total oxidized N (ppm)	2.21
Alkalinity (ppm CaCO3)	125.00
Chloride (ppm)	29.10

(b) *Invertebrate groups (referred to by number) predicted from 11 physical and chemical variables:*
44 —49.9% 105—17.2% 23—15.7% 51—10.8% 104—3.5% 32—1.5%

(c) *Selection of predicted taxa, in decreasing order of probability:*
99.4% *Micropsectra/Tanytarsus*
89.7% *Limnodrilus hoffmeisteri*
79.8% *Dicranota* sp.
69.1% *Baetis scambus* group
59.6% *Asellus aquaticus*

(d) *Observed number of taxa:*
Expected: at 75% = 0.972; at 50% = 1.017

Source: Wright *et al.* (1989).

Plate 7.2. Electro-fishing survey of a small upland stream.

sensitivity. However, elsewhere we have also referred to the robustness of the freshwater environment, particularly of river *systems*. We now need to be more circumspect about how various forms of impact alter the basic controls on freshwater habitat; the concept of *stress* will be more helpful than one of *damage* (just as in Chapter 8 we carefully separate contamination from pollution).

The mechanisms by which manipulations of the environment cause stress to organisms are shown in Box 7.3. Clearly, when one member of an ecosystem becomes stressed in one of the ways listed, the complex interrelationships of the system permit both a measure of adaptive survival strategies and a system-wide vulnerability. One may consider the former to operate up to a point at which adaptability can no longer defend the system—a threshold point (see Section 4.3).

We restrict this section to rivers and lakes because wetlands are far more susceptible to physical impacts than they appear to be to pollution (Section 7.7).

7.6.1 Pollution

We have mentioned in passing that the chemical content of water is important to freshwater life in controlling:

- cellular balances and exchanges;
- nutrition;
- respiration.

Just as humans once complained of the effects on health of 'a change of water' (before uniform standards reduced local and regional variability), any change in the ionic content of fresh waters will bring about an organic response. Natural fluctuations abound, however, in the seasonality of climate or episodic nature of precipitation (see Section 8.1.1).

Ecotoxicology, a branch of chemical toxicology (Section 8.2), attempts to establish the pattern of stress produced on organisms in relation not only to their own bodily functions but in an ecological context of community structures, food chains, predation, etc. Food chains are a potent force in exacerbating the damage from freshwater pollution incidents, both by producing *bioconcentration* up the chain as the toxins accumulate and in spreading the impact through the system (and therefore increasing its duration).

Consideration of water pollution is always dominated by chemical considerations but sediment pollution is receiving much more research attention in an era of ambitious land/soil management by farmers and foresters. The dangers of high concentrations of suspended solids in rivers have been respected since the first sewerage regulations of a century ago. Industry is capable of producing effluent with a high solids load, as is quarrying.

Table 7.7 summarizes the main impacts on fresh waters of 'conventional' point sources of pollution. Increasingly the predictable outcomes of this relationship have become the basis of a biological-survey approach to policing water pollution control laws. Extence and Ferguson (1989) reveal the practicability of invertebrate surveys in regulatory systems; it is of particular benefit that the biological record of pollution incidents is often available (as dead organisms or 'missing' species or families) long after a chemical test could reveal the responsible pollutants. See Chapter 8 for a consideration of water pollution at a much more detailed level.

7.6.2 River regulation

Because the regulation of rivers deploys stored water to reduce the natural extremes of flow there are ecological stresses on both lake/reservoir and downstream biota; most research has focused on the latter.

Petts (1984) encourages the synoptic view of downstream efects as comprising:

- First-order impacts:
  flow regime;
  water quality/sediment load;
  physical barrier/plankton;

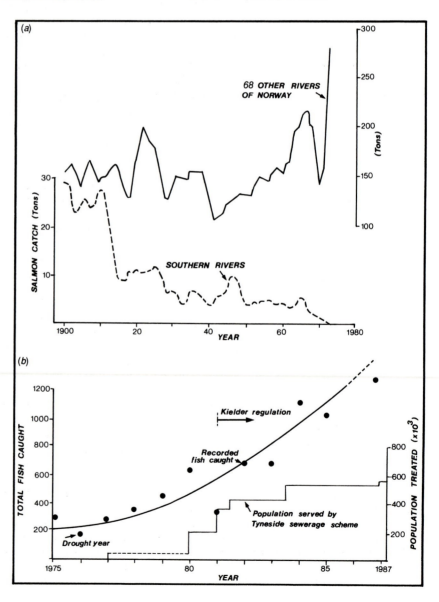

Fig. 7.10. Fish as indicators of water-quality decline and recovery. (a) Indications of acidification damage in southern Norway. (b) The recovery of the River Tyne, England, after sewerage improvements and river regulation to improve water quality (after Johnson, 1988).

Box 7.3. *The trends that may be expected in ecosystems upon the advent of stress*

Energetics:
1. Community respiration increases.
2. Production/respiration becomes unbalanced (i.e. P/R becomes greater than or less than 1).
3. P/B and R/B (i.e. maintenance to biomass structure) ratios increase.
4. Importance of auxiliary energy increases.
5. Exported or unused primary production increases.

Nutrient cycling:
6. Nutrient turnover increases.
7. Horizontal transport increases and vertical cycling of nutrients decreases.
8. Nutrient loss increases (system becomes more 'leaky').

Community structure:
9. Proportion of r-strategists increases.
10. Size of organisms decreases.
11. Life spans of organisms or parts (leaves, for example) decrease.
12. Food chains shorten because of reduced energy flow at higher trophic levels and/or greater sensitivity of predators to stress.
13. Species diversity decreases and dominance increases; if original diversity is low, the reverse may occur; at the ecosystem level, redundancy of parallel processes declines.

General system-level trends
14. Ecosystem becomes more open (i.e. inputs and outputs become more important as internal cycling is reduced).
15. Successional trends reverse (succession reverts to earlier stages).
16. Efficiency of resource use decreases.
17. Parasitism and other negative interactions increase, and mutualism and other positive interactions decrease.
18. Functional properties (such as community metabolism) may be more robust (homeostatic-resistant to stressors) than are species composition and other structural properties. In systems dominated by long-lived perennial plants (e.g. forests), the reverse may be true.

Source: Freedman (1989).

• Second-order impacts (both indirect and delayed):
   channel form/substrate;
   macrophytes/periphyton;
• Third-order impacts:
   invertebrates/fish.

He indicates that third-order impacts will reflect all the changes of the two more direct categories, together with those of the adjacent terrestrial environment, e.g. as a result of floodplain development following regulation and irrigation/power generation.

We have already addressed several aspects of the

effects of regulation (see Section 5.8.4); the impacts on channel and corridor habitats are becoming more salient to discussions of river management as the biological database grows; certainly the third-order impacts are now part of environmental-impact assessment for large water resource projects (Canter, 1985).

Included under the heading of river regulation is the use of *water transfers* to make maximum benefits of river-channel conveyance to sites of water use or power generation. Section 7.5 demonstrated the ecological relationships between plant and animal communities and types of river channel. The stress on those communities introduced by water transfers may be dominated by the addition to the receiving stream of water of unusual chemical quality but of course there is also a danger of transferring biota themselves. The transfer of coarse-fish predators to a game-fish river or lake would clearly be disastrous but there are many more subtle transfer dangers including that of disease.

7.6.3 Channelization

Flood protection for residential, industrial, and agricultural land normally requires modifications to the width, depth, gradient, or planform of natural stream channels; the North American term for such processes is *channelization*, though a European alternative is *river training*. As we reveal in Section 7.7, flood protection often leads to loss of wetlands, both directly on floodplains and because it often accompanies land drainage.

Brookes (1988) has compiled a major review of the techniques and impacts of channelization. In terms of river biota Brookes considers both the direct effects of a profound change in the substrate and velocity conditions of an engineered reach and the indirect effects of downstream physical disturbance and changes in the flow regime. Initial impacts are often more spectacular on invertebrates and fish than are the longer term effects as the new channel naturalizes and recruitment/annual migration/drift restores the former conditions. Impacts extend beyond the channel to the river corridor, particularly if this is cleared for access by machines.

Many countries have now introduced standards for the protection of rivers undergoing channelization. In the UK *environmental-impact assessment* is required before major schemes are carried out on natural channels; project appraisal has become widespread and includes close attention to issues of nature conservation (Holmes, 1991).

Table 7.7. Environmental implications of the discharge of sewage and industrial effluents

Factor	Principal environmental effect	Potential ecological consequences	Probable severity	Remedial or ameliorative action	Comments
Organic enrichment					
High biochemical oxygen demand (BOD) caused by bacterial breakdown of organic matter	Reduction in dissolved oxygen concentration	Elimination of sensitive oxygen-dependent species; increase in some tolerant species: change in community structure	Dependent upon degree of de-oxygenation, often very severe	Pretreatment of effluent; ensure adequate dilution	BOD can be reduced substantially by adequate treatment of effluent before discharge
Partial biodegradation of proteins and other nitrogenous material	Elevated ammonia concentrations; increased nitrate levels	Elimination of intolerant species since ammonia is toxic: reduction in sensitive species; potential for increased plant growth in nutrient-poor waters	Variable, locally severe Mild/moderate	Provision of improved treatment to ensure complete nitrification: nutrient stripping possible but expensive	As above, adequate treatment is best solution to this problem; de-nitrification is the ultimate solution
Release of suspended solid matter	Increased turbidity and reduction of light penetration	Reduced photosynthetic activity of submerged plants: abrasion of gills or interference with normal feeding behaviour (see inert solids below)	Moderate usually local	Provide improved settlement, ensure adequate dilution	
Deposition of organic sludges in slower water	Release of methane and hydrogen as sulphide matter decomposes anoxically	Elimination of normal benthic community	Variable, may be severe	Discharge where velocity adequate to prevent deposition	This tends to be a locally restricted phenomenon
	Modification of substratum by blanket of sludge	Loss of interstitial species: increase in species able to exploit increased food source			

Toxic wastes				
Presence of poisonous substances	Change in water quality	Water directly and acutely toxic to some organisms, causing change in community compositions: consequential effects on prey–predator relations; sub-lethal effects on some species (impaired reproductive capacity, changes in behaviour, etc.)	Highly variable, depending upon substance and its concentration	Little can be done except to provide increased dilution	Toxic effluents cover a very wide range of substances and it is, therefore, difficult to generalize
Inert solids					
Particles in suspension	Increased turbidity Possibly increased abrasive action	Reduced photosynthesis of plants. Impaired feeding ability through reduced vision or interference with collecting mechanisms of filter feeders (including abrasion or reduction in nutritive value of collected material).	Variable, moderate	Provide improved settlement facility	Inert solids may cause greater change than organic wastes since, although they change the character of the substrate and are unstable, provide no additional nutrition
Deposition of material	Blanketing of substratum, filling of interstices and/or substrate instability	Change in benthic community, loss of interstitial species, reduction in diversity, increased number of a few adventitious species, substrate is unstable.	Variable, often severe	Discharge where velocity adequate to ensure dispersion	

Note: The effects of the three major categories of effluents, namely degradable organic matter, toxic substances, and inert solid particles, are considered separately. Many effluents are composed of more than one type and the proportions of these vary according to the source.

Source: Hellawell (1986).

Plate 7.3. Engineered rural channel showing compound section between flood embankments. In the middle distance is a rated-section gauging-station.

Since the techniques of channelization include the dumping of spoil on river banks or floodplains, or the construction of modified river banks (Plate 7.3) it has been an important success by conservationists to extend the assessment of impacts to a river corridor outside the channel itself. Whilst conservation duties of river managers may include physiographic and architectural features, it is the more easily observed and recorded plants and birds that form the cornerstone of corridor surveys performed in England and Wales (see Fig. 7.11). Despite the success of the policy to protect river-bank habitats it is important to note that regular riparian *management* is still a requirement if ageing and damaged trees are not to block bridges during floods (Plate 7.4).

Holmes's system of classification has become the basis of assessing the conservation status (typicality, rarity) of channel vegetation for use by British river-management authorities and conservation bodies; it also helps form the basis of repeatable surveys of the expansion/contraction of species (in geographical range).

7.7 Wetland systems: Saturated and inundated habitats

We have observed that the fresh waters of lake and river systems cover barely 2 per cent of the land surface in a humid, oceanic climate like that of Britain. Wetlands by contrast occupy 6 per cent of the Earth's surface; they result from the interaction of the *water-table* (very broadly defined as the general level of saturation) with the *land surface*. The water-table oscillates over long time periods according to gross environmental change, as the record in peat bogs demonstrates, but also seasonally in many climates.

Wetlands also control their own hydrological relationships by growing or declining in surface level relative to the water-table or open water bodies. The classic *hydrosere* sequence of vegetation and wetland geometry which has operated in landscapes previously glaciated is shown in Fig. 7.12. The sequence shown is highly simplified but it illustrates a warning to environmental scientists and conservationists that, since wetlands are a dynamic feature of landscapes, changes in their status *may not always originate from an artificial cause*. The figure also illustrates the importance of the source of supply to the water-table of a wetland runoff to a lake, seepage from groundwater to a fen, and rainfall to a raised bog.

7.7.1 Wetland classifications

Research interest in wetlands has mainly been the province of biologists; botanists and palaeobotanists (interested in the plant record the peat wetlands preserve) have only recently been joined in their

Fig. 7.11. A typical river-corridor survey carried out in England and Wales prior to engineering activity on rivers.

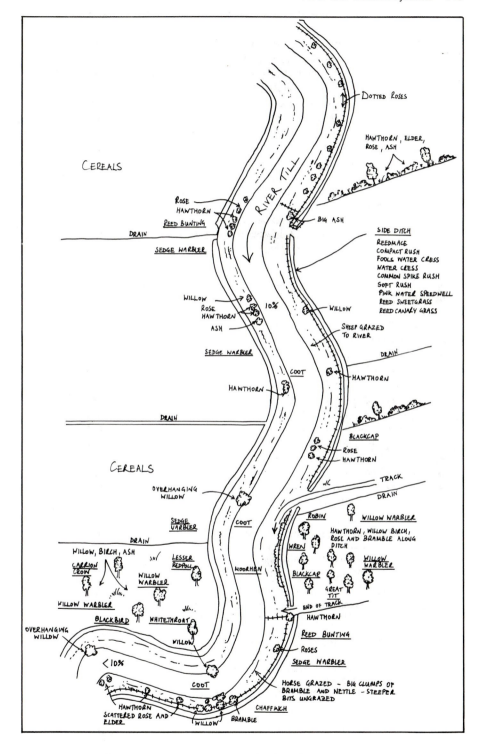

Plate 7.4. A bridge over the river Wansbeck at Morpeth, Northumberland dangerously blocked by fallen timber from the riparian zone upstream.

endeavours by geomorphologists and hydrologists. Wetlands are becoming a prime focus for interdisciplinary research efforts.

Fig. 7.12 has already introduced a topographic, cross-sectional view of some major wetland types. The relationship between topography and water supply means that, at the extremes, a fen will be supplied by nutritionally 'rich' seepage waters with a high solute content, including nutrients, whilst raised bogs are supplied by 'poor' rainfall. The terms *eutrophic* and *oligotrophic* are used for these cases, with *mesotrophic* conditions representing the intermediate case.

The fine detail of the botanical composition of wetlands has been a feature of most systems of classification until recently when the specialized knowledge required to apply them became a popular application in the interests of conservation. Box 7.4 shows alternatives, of which topographic systems are increasingly popular (Ratcliffe and Oswald, 1988) as are hydrological schemes (Box 7.5).

For the specific purposes of assessing the threats to conservation management of wetlands Newson (1992) proposed the simple hydrological classification of British wetlands shown in Box 7.5. Size is by no means a minor consideration when one considers that most threats to the saturation of a bog will work inwards from the periphery, e.g. drainage of adjacent agricultural land.

Other hydrological classifications tend to be developments of that shown in Fig. 7.12 but without its evolutionary sequence, utilizing instead the variety of sites for wetlands in the landscape such as that shown in Fig. 7.1(a). Clearly the maintenance of

Box 7.4. *Methods of mire classification*	
Morphological:	e.g. raised, domed, etc.
Ecological:	e.g. topogenous, soligenous, ombrogenous—sources of inflow
Topographic:	e.g. plateau, saddle, basin, valley (hydrologically useful)
Vegetational:	Traditionally by plant community; more recently multivariate statistical classification
Palaeobotanical:	e.g. sequence of evolution revealed by pollen analysis
Chemical: eutrophic	e.g. oligotrophic, mesotrophic, or base-poor/base-rich
Exploitational:	e.g. calorific value of peat, horticultural workability/nutrition
Source: Ratcliffe and Oswald (1988).	

wetlands in a fluvial landscape moderates and modulates the extremes of both flow and pollution by acting as a storage and exchange site; loss of wetlands almost always occurs during economic development in favour of landscapes and land uses which reverse this situation, producing greater extremes of runoff (e.g. urbanization) and/or greater pollution (e.g. intensive agriculture). This point is returned to in Section 7.10.

7.7.2 Wetland substrates and hydrological processes

Whilst, under the simple definitions of wetlands provided above, it would be possible for a basin filled with glass balls to act as a wetland, i.e. to be permanently or seasonally saturated and/or inundated,

Fig. 7.12.
Development
stages from
post-glacial lake
to raised bog,
showing main
sources of
wetland
recharge.

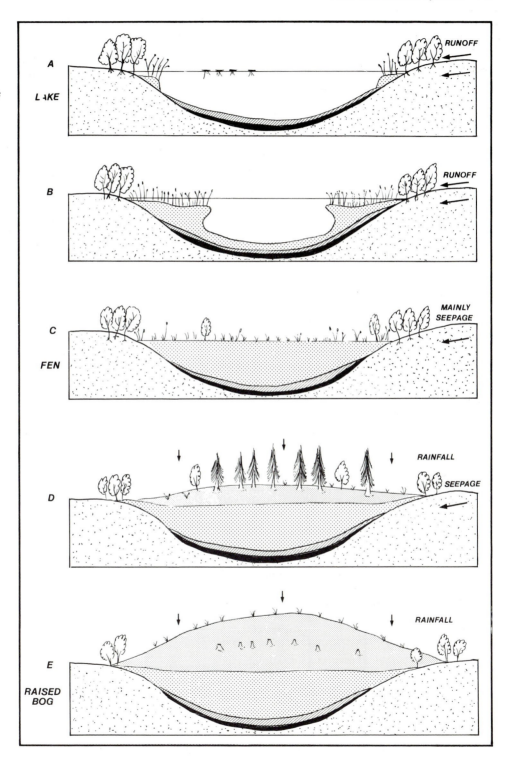

Box 7.5. *A simplified classification scheme for UK wetlands*			
Source of water	Extent		
	Small <50	Medium 50–1000	Large >1000 ha
Rain	R1 Parts of some basin mires	R2 Raised mires	R3 Blanket mire
Springs	S1 Flushes	S2 Fen basins	S3 The Fens, Somerset Levels and Moors
	Acid valley and basin mires		
Floods	F1 Narrow floodplains	F2 Ings, typical floodplains	F3 Washes, E. Anglian marshes, Somerset Levels and Moors

Source: Newson (1992).

the existence of blanket bogs and raised bogs indicates that the hydrology of wetland substrates plays a major part. Poor permeability (or, more correctly, *low hydraulic conductivity*) ensures that the peat and clay soils which dominate wetlands have an innate ability to store water, even in periods when the input/output balance does not suggest saturated conditions.

In Chapter 3 the brief discussion of soil hydrology emphasized the large range of values adopted by hydraulic conductivity. Clay soils are at the extremity of most conventional tabulations (see Fig. 3.5) but within organic soils peats have a very wide spread of conductivities. Peat is formed by the very slow decomposition of plant material under the anaerobic and often acidic conditions of the wetland surface. The process of *humification* can be measured qualitatively in the field by simple observations of the consistency, fibre, content, and moisture content; the best-known scale of such measurements is that by von Post and it is possible to broadly link the von Post scale to hydraulic conductivity (Fig. 7.13(a)). The coarser-fibred fen peats are clearly more permeable at a given humification than the bog peats (e.g. those of raised or blanket bogs).

One of the most intriguing features of both clay and peat wetland soils in the field is the variability encountered in their generally low rate of throughput of soil water. In fact we have already observed a deal of heterogeneity in both clays (cracking can

control the success of drainage: see Section 5.5.2), and peats (in the case of piping phenomena covered in Section 3.7.4). Further details of the heterogeneity of peats have recently emerged from a succession of studies, notably by Ingram (1982) of the *diplotelmic* structures of the upper layers of peat bogs, notably raised bogs. The diplotelmic term refers to the relatively sharp vertical zonation into an upper 10 cm or so of relatively fibrous, permeable peat (*acrotelm*) and an underlying humified, impermeable mass of peat (*catotelm*) through which cracking, piping, or artificial drainage may provide the only relatively rapid water transmission. The diplotelmic form of peat stratification is critical to the shape and function of raised bogs.

Ingram (1982) suggests that it is not capillary forces (see Sections 2.7 and 3.4) which maintain water at a high level, close to the surface of raised mires; instead he proposes that the body of water which saturates the catotelm is a *groundwater mound* representing the dynamic equilibrium between recharge from rainfall and extremely slow lateral seepage. If the hypothesis is true it should be possible to calculate the surface profile of the mound by the following equation (derived from Darcy's law):

$$\frac{U}{K} = \frac{H^2}{2Lx - x^2} \quad \text{(dimensions indicated on Fig. 7.13(b)).}$$

This predicts the shape of an ellipse for the raised mire water-table (Fig. 7.13(b)). The maximum height of the mire (Hm) is given by

$$\frac{U}{K} = \frac{Hm^2}{L^2}$$

By gathering figures for a site by field-work (hydraulic conductivity) and inspection of records (rainfall, evaporation) it is therefore possible to calculate the relative hydrological 'health' of a raised mire, or to predict the shape eventually taken up by the water-table in a damaged mire.

Ingram himself performed detailed field measurements at a raised-mire site in Scotland (Dun Moss) which is largely undamaged; a good agreement with the theoretical ellipse was obtained (Fig. 7.13(b)).

7.7.3 Vegetative feedbacks: evaporative loss and interception

Chapter 2 dealt with the processes which power evaporation and the way in which plant-canopy structure and physiological processes control actual rates of both evaporation and transpiration. Because of our evolutionary concept of some major wetland types and the hydrosere's progressive change in plant-cover types it is only to be expected that there will be a variety of feedback processes operating in wetland near-surface hydrology. These feedbacks have, however, proved difficult to research.

For example, very few field studies of wetland hydrology have measured the interception effects of the thick, seasonal canopy of intermediate-height plants which typically cloaks temperate wetlands (Plate 7.5(a)). Interception rates of up to 40 per cent of incoming rainfall are possible on an annual basis, though measurement of stemflow is problematic for such a dense crop cover.

It is often assumed that, under saturated conditions, because wetland plants are not limited for water uptake they display very high rates of transpiration—in excess of 10 mm per day; indeed the common reed, *Phragmites australis*, has been termed 'a water pump' because of its alleged capacity to dry out the wetland surface. However, as discussed in the case of forest hydrometeorology, edge effects are very influential in controlling wetland transpiration rates; it is essential, therefore that measurements are made well into the transpiring vegetation body (see Plate 7.5(b)).

7.8 Stressed wetland environments and protection

In Section 7.6 a contrast was set up between drainage to freshwater biota, via habitats, and stress which can have multiple contributory factors. The situation is simpler but more sensitive in wetlands for the following reasons:

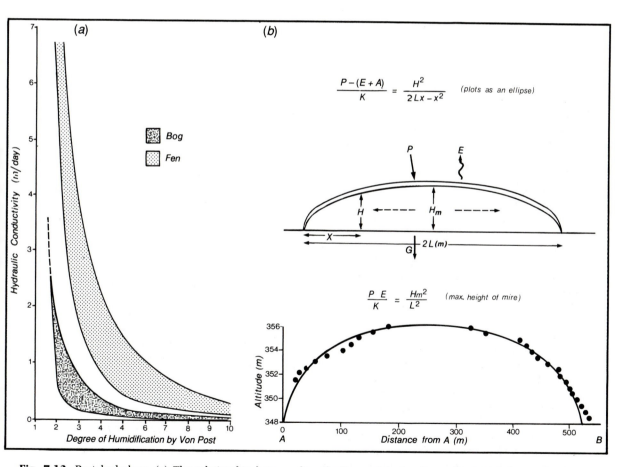

Fig. 7.13. Peat hydrology. (a) The relationship between humification and hydraulic conductivity (after Castle *et al.*, 1984). (b) Ingram's model of the groundwater mound which forms the characteristic profile of raised bogs (after Ingram, 1982).

• Plants dominate the basic conservation interest, forming the major habitat for other species worthy of conservation; *plants can be destroyed* in a short time or fail to reproduce after a single season of moisture or water-quality stress.

• The threat to wetlands is often *direct* and *total* (as in reclamation), not accidental or incidental as in the case of threats to rivers and lakes.

• Where the threat is peripheral and damage is peripheral that *damage is progressive* and increases the likelihood of further threat as degradation of habitat works towards the wetland core (unless buffer zones or physical obstructions such as dams or membranes are installed).

However, as in the case of rivers and lakes it should be remembered that some stress is natural and the characteristic biota are generally adapted to it (e.g. in regulating transpiration, in adopting microtopographic zonation, and adopting vegetative reproduction/dormancy).

7.8.1 Drainage

Many of the world's most fertile and most commercially advantageous sites would not have been set-tled by mankind without drainage to prevent seasonal or permanent saturation and/or flood protection to reduce inundation. Table 7.8 therefore catalogues achievement or disaster depending upon the reader's viewpoint!

Table 7.8. Percentage of agricultural land drained by country

Country	Percentage of agricultural land
Northern France	10.4
France (total)	10
UK	60.9
Germany	37.1
Italy	24.2
Belgium	22.5
Netherlands	65.2

Note: These figures may not be wholly accurate and they do not allow straightforward comparisons. There are major differences.

Source: Baldock (1984).

Conventionally the 'battle for bogs' is a lowland phenomenon, with fens and basin or flood wetlands prominent because of their agricultural value (they are nutritionally rich). Plate 7.6(a) illustrates a

Plate 7.5. (a) The wetland (fen) vegetation canopy, looking upwards from ground level. (b) Use of micrometeorological measurements and weighing lysimeters to index wetland evapotranspiration.

Plate 7.6. (a) A typical wetland nature reserve boundary: fence, ditch, and 'productive' agriculture. (b) The typical topography of the Flow Country of Scotland, UK. (c) Peat extraction from a lowland raised mire: Thorne Moor, UK.

fairly typical boundary between the competing land uses; the erection of a fence to mark the boundary of a wetland reserve is of course no protection against peripheral drainage impacts. However, in many developed agricultural nations the rival claims of wetland conservation and improved agriculture or forestry now also clash in remote areas such as the Flow Country of Scotland (Ratcliffe and Oswald, 1988). As we saw in Chapter 5, the full aims of drainage are seldom attained in peats of low permeability with the result that when drainage is carried out a wetland is spoiled by a laceration of channels but the land is very little improved agriculturally. In the Flow Country a very distinctive surface topography of open pools is also under threat from drainage (Plate 7.6(b)). Another ele-

ment of the 'battle for bogs', peat extraction (Plate 7.6(c)) has recently been addressed by a vigorous protest campaign in the United Kingdom; a major result has been the introduction of peat substitutes in horticulture.

In a study (Gilman and Newson, 1982) of two fen mires in Anglesey threatened by agricultural drainage, the Institute of Hydrology established comprehensive hydrological measurement networks designed with two objectives:

(a) To calibrate the gross recharge and loss of each system to assess the hydrological viability of the fens (asssuming water quality is suitable).

(b) To investigate the 'preferred' peat water levels of the conservationally desirable plant species (principally saw sedge—*Cladium mariscus*). This was the reason for selecting two sites—one being undamaged and the other clearly losing much of its *Cladium* through drying out.

Because the aim of the study was to improve the protection of the damaged fen—Cors Erddreiniog—the illustrations here refer to that site. They indicate (Fig. 7.14) the hydrological network and gross water balance, but also a series of three options for

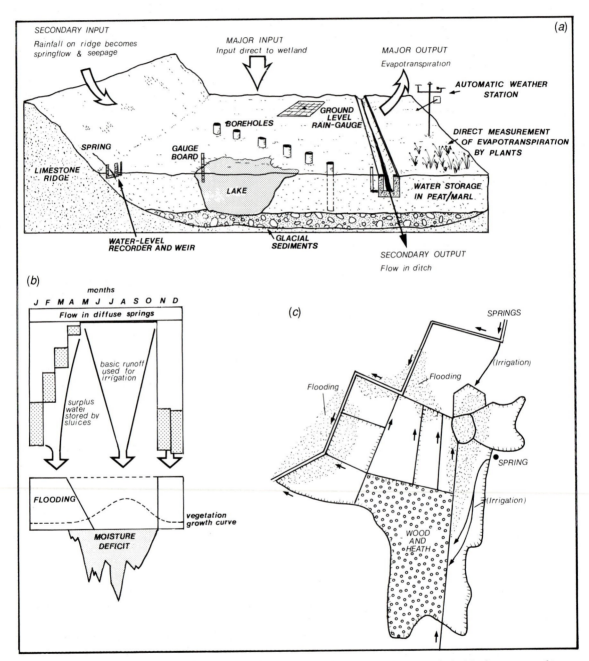

Fig. 7.14. Practical hydrology on a threatened wetland (Cors Erddreiniog, Anglesey, Wales). (a) The pattern of instrumentation in relation to the fen basin. (b) The annual requirement for 'irrigation' as a water deficit develops and the wetland lacks surface flooding. (c) A simple management plan to re-wet damaged parts of the wetland with water of appropriate volume and quality (all after Gilman and Newson, 1982).

retaining the high water levels typical of winter and early spring. The main stress for *Cladium* seemed to derive from dry spring seasons in which other species competed against the sedge. The small-scale proposal was selected because of the problems of neighbouring landowners who required (and had secured) efficient drainage. It is important, too, to note that such small-scale wetland management schemes are within the resource and manpower limits of volunteer forces, involving the construction of small dams and the excavation of feeder ditches.

It is critical at Cors Erddreiniog to ensure maximum retention of base-poor rainfall/surface flows to create surface saturation/inundation but also to distribute the base-rich water from the Carboniferous limestone as widely as possible to retain fen-peat conditions. Reduction of both quantity and quality of recharge waters seriously threatens all wetlands in different ways.

On the raised mires of oceanic climates, damage by drainage or the plantation of forests may not only be a direct impact. Smith and Charman (1988) identified the drying influence of conifer forests in northern England by comparing the botanical composition of thirty-four upland mires completely surrounded by forest with features such as forest age and mire geometry. Mires with a long forest boundary and with trees older than twenty years were more likely to have dry moorland species of plant as opposed to undisturbed ombrogenous mire species. There is thus circumstantial evidence that the forest influence extends beyond the plantation, either by creating some form of rain-shadow or, more likely, by lowering water-tables within the plantation. In peat soils, dominant in the study area, interception causes shrinkage of peat accentuating both the surface slope off the mire margin and the gradient of the water-table.

7.9 Material benefits of the freshwater environment

'Wetlands are not wastelands' proclaims an editorial note in *Water Quality Bulletin* (not a conservation magazine!). President George Bush of the USA stated 'My position on wetlands is straightforward: all existing wetlands, no matter how small, should be preserved' (*Sports Afield Magazine*, October, 1988). Conservationists were therefore horrified by ex-President Bush's later attempts to redefine wetlands so that some could in fact be sacrificed to development!

We are witnessing, at least in developed societies, a relatively rapid reversal of an attitude which has lasted since mankind first invented a plough with which to cultivate and drain. Whilst early modern communities had considerable respect for clean rivers (see Parker, 1976) and for the economy of seasonal floodlands (see Darby, 1983), the Industrial Revolution in developed nations viewed rivers as sewers and wetlands as either granaries or dumps.

Attitudes are now changing. The direct value of wetlands for conservation and amenity is being quantified; it is clear that a passionate interest in wetlands as natural features extends well beyond the technical enthusiasts (such as Britain's David Bellamy). Society is beginning to learn, through research and monitoring that wetlands both reduce the river-flood hazard and help to purify polluted waters (Table 7.9).

Table 7.9. Treatment performance of an artificial wetland system on Long Island, NY (in mg l^{-1})

Measure of water quality	Water flowing into wetland	Water flowing out of wetland
Biochemical oxygen demand	520	16
Suspended solids	860	57
Nitrogen	36	4
Faecal coliform bacteria count per 100 ml	3000	21

Source: Modified from data of Small (1972); report by Fritz and Helle (1978).

One of the overlooked but unquantifiable benefits of peatlands is the record of past environmental changes which they preserve; because our instrumental records of changes in climate and runoff are very short (*c*.200 years) we are unable to make mathematical predictions of future change with much precision but the peat record illustrates the range within which changes might occur and their rapidity.

Finally, the wet and wet*land* environments covered by this chapter are belatedly being seen in the integrated fashion advocated by the early sections of this chapter. It is now quite feasible to envisage planted reed-beds used as 'natural' sewage works in the headwaters of a stream whose floodplains and fens have been conserved as buffers to nutrient or sediment runoff, flowing into a protected lake used to store flood waters.

7.10 Tropical/semi-arid wetlands and development threats

Tropical wetlands are amongst the most productive ecosystems on Earth but are much threatened by opposing claims of drainage for agriculture and flooding for water storage. Because many of them are in coastal locations they are brackish and sport mangrove forests. They are important, as are most tropical ecosystems, for their huge biodiversity and therefore conservationists are attempting to prevent destructive developments. In some cases, development is inevitable and attempts are urgently being made to make this sustainable.

For example, in Jamaica, Bjork and Digerfeldt (1991) recommend peat extraction as a development process because the open-water areas which result can then play their part in shrimp fishing as a human livelihood and the attraction of wildlife (if they are not invaded by water hyacinth). One unexpected saving grace for the swamp forests of Jamaica is the government's refusal to allow them to be felled if the replacement crop is cannabis for the drug trade! Predicted global sea-level rise will also threaten coastal wetlands, and whilst the Jamaican systems kept pace with annual rates of rise of 3.8 mm during the Holocene, their conserva-tion in the face of 'greenhouse' rates of rise is extremely risky.

Hollis (1990) itemizes the threats to water quantity, water quality, and traditional sustainable exploitation in arid and semi-arid wetlands. There are particular threats to the latter around the Mediterranean lands: Tunisia, Algeria, Greece, and Spain have particular problems associated with irrigation and tourism development. In Tunisia, Lake Ichkeul, a 90 km^2 seasonally saline lake, surrounded by 30 km^2 of freshwater wetlands, is critically important for overwintering waterfowl. Three dams on inflowing streams have been built as part of an irrigation scheme; the reduced inflows in winter will render Ichkeul saline all year and dry out its wetlands. The Mikri Prespa wetland in Greece has been threatened by diversion of the inflowing streams to feed irrigation schemes, by channelization, by the prohibition of reed cutting, and by commercial fish farming.

On the Atlantic coast of south-west Spain the Coto de Doñana is a 770 km^2 system of sand dunes, heathlands and wetlands—the marismas—which are critically dependent on winter recharge. The first signs of trouble occurred when pesticide pollution of the inflowing Guadalquivir River reduced wildfowl numbers (see Stevenson, 1992).

8. Water pollution

There would be good grounds for placing this chapter much earlier in the book; we have referred to pollution on many occasions. However, most readers will have grown up in circumstances where pollution as a general principle is well understood and previous references here have not worked outside the kind of general knowledge gained from media coverage of local and global problems caused by pollution of fresh waters. It is now appropriate to go into more detail.

8.1 Definitions: source, pathway, target

We need to develop a more formal treatment of water pollution related to our knowledge of hydrological pathways and processes. The definition of pollutants offered by Holdgate (1979) is a popular one, especially amongst geographers, focusing as it does upon substances causing *damage* to *targets*, with emissions from *sources* into *pathways* in the environment along which the pollutant reaches the target.

The geographical element is clearly in the tripartite division of sources (which have locations), targets (which also have locations), and pathways (which imply travel between and within locations). Pollution and pollution control exhibit, therefore, the inherent *space/time frameworks* enjoyed by geographers (Newson, 1991). Pollution has patterns and processes which are open to geographical treatments: it can be mapped and modelled, although much of our work is inevitably partial (e.g. investigating, through health records, the distribution of targets for water-borne disease from sewage). The pathways in the hydrological system have been laid bare by the seven preceding chapters and are developed for this chapter in Fig. 8.1; we now need to show how they operate.

First, however, it is necessary to define water pollution against a natural standard and to compare it with other standard terms such as contamination.

8.1.1 What is 'pure' water?

The need for care in defining water pollution is well illustrated from the field of *surface-water acidification*, better known as 'acid rain'. The fact is that 'natural' rainfall is acid in reaction, averaging pH 5.6 (against a neutral reaction of pH 7.0). Rainfall is contaminated by gases of the Earth's atmosphere including carbon dioxide and sulphur dioxide emitted, for example, by volcanoes.

In any study of terrestrial water quality, therefore, *inputs from precipitation* cannot be discounted. The acidification issue (see Section 8.6) has educated us to the fact that atmospheric pollution can become freshwater pollution through rain-out and through *dry deposition* on surfaces such as those of vegetation canopies (followed by washing-off during precipitation).

The atmospheric contribution to precipitation chemistry is indicated by the analyses in Table 8.1(a); sampling is very problematic, particularly if one wishes to separate the dry deposition from that actually dissolved in rain or snow as it falls. Further problems arise from bird droppings or other local sources of contamination. Nevertheless broad patterns emerge in a climatologically stratified land mass such as Britain with westerly stations (such as Pwllpeiran in Table 8.1(a)) illustrating the influence of prevailing winds across the Atlantic Ocean (e.g. in sodium and chloride inputs), and easterly stations (e.g. Boxworth) showing more nutrients and metals. An additional contrast exists in the total input of precipitation; thus the volumes of precipitation collected at Pwllpeiran far exceed those at Boxford and therefore the *loadings* of solutes may reach higher totals at the wetter site despite lower *concentrations*.

Hydrological pathways broadly quantify the ionic composition of sources used for water supply; Table 8.1(b) shows that upland streams are more dilute than lowland streams, though more acid. Surface runoff routes mean that there is little chance for solution of biogeochemicals, compared with the infiltrating waters of the drier, thicker soils of the lowland river catchment. Groundwaters become

Fig. 8.1. A repeat of Fig. 5.1 with the runoff-process 'tanks' related to contamination/pollution.

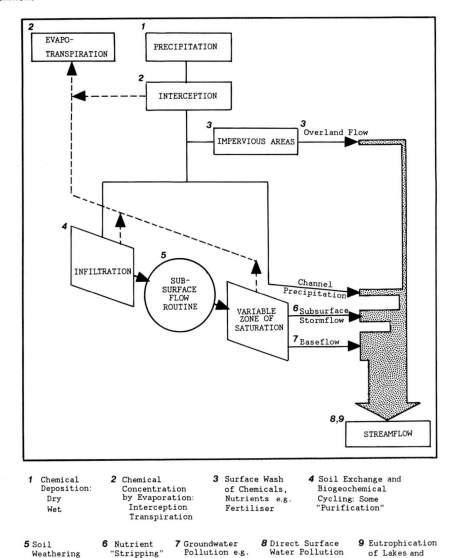

1 Chemical Deposition: Dry Wet	2 Chemical Concentration by Evaporation: Interception Transpiration	3 Surface Wash of Chemicals, Nutrients e.g. Fertiliser	4 Soil Exchange and Biogeochemical Cycling: Some "Purification"
5 Soil Weathering and Chemical Leaching	6 Nutrient "Stripping" Exchange	7 Groundwater Pollution e.g. Waste Disposal	8 Direct Surface Water Pollution from Point Sources

9 Eutrophication of Lakes and Seas

highly concentrated with solutes because of the weathering processes occurring during percolation.

8.1.2 Contaminants, pollutants, organic threats

The definition we have used for pollution (Holdgate, 1979) refers to harm/damage being done to the targets involved. There is clearly a threshold between concentrations of substances which are harmless but above the natural background of pristine environments (which constitute *contamination*) and those which cause harm (*pollution*). Toxicological studies (see Section 8.2) have revealed this threshold value for but a few major substances.

The major contaminants of surface waters are shown in Table 8.2(a); in most cases specific *concentrations* or *loadings* (total inputs over time) are sufficient to cause pollution. In developed nations the corollary of high levels of output of contaminants is a regularized system of collection, processing, and discharge of waste known as *sewerage* (the pipe network and works) carrying *sewage* (the wastes). Table 8.2(b) shows how large a volume of discharge arises for various sources in the United Kingdom; treated sewage discharges can become a major component of river flow in some locations, particularly during droughts (Plate 8.1). The principle of *sewage treatment works* is to purify the effluent

Table 8.1(a). Mean values of mineral content of rainwater for summer and winter rainfall, 1968 and 1969–1972

	Pwllpeiran (Wales)		Boxworth (Cambs.)	
	Summer	Winter	Summer	Winter
Volume (l)	48.37	58.26	20.43	18.62
Insoluble matter (mg)	155	140	439	307
Ash (mg)	85	66	165	186
Ash (as % insoluble matter)	55	47	38	61
Mineral content (mg/l)				
Ca (Calcium)	0.49	0.25	1.96	1.32
Mg (Magnesium)	0.23	0.34	0.19	0.26
K (Potassium)	0.12	0.14	0.54	0.39
Na (Sodium)	1.19	2.49	1.32	2.16
NH_4-N (Ammonium Nitrogen)	0.22	0.16	1.52	1.29
NO_3-N (Nitrate Nitrogen)	0.13	0.12	0.75	0.70
C1 (Chlorine)	3.11	5.31	4.20	4.90
SO_4-S (Sulphate Sulphur)	1.02	0.89	2.53	2.28
Cu (Copper)	18	17	10	7
Zn (Zinc)	164	164	379	487
P (Phosphorus)	37	15	94	53
F (Fluorine)	13	9	87	95
B (Boron)	9	6	28	28
Co (Carbon Monoxide)	0.1	0.1	0.5	0.5
Mo Molybdenum	0.1	0.1	0.8	2.7
Pb (Lead)	6.5	3.6	7.9	12.0
Sn (Tin)	0.1	0.1	0.2	0.5
V (Vanadium)	1.1	0.5	2.0	3.7
Cr (Chromium)	0.4	0.2	1.9	2.0
Ni (Nickel)	1.2	1.4	5.6	3.6
Ti (Titanium)	0.3	0.4	1.1	2.0
Mn (Manganese)	5.8	3.6	22.7	17.5

Source: Modified from Wadsworth and Webber (1980).

Table 8.1(b). Characteristics of various water sources

Characteristic, mg/l (except where noted)	Source		
	Upland catchment	Lowland river	Chalk aquifer
pH (units)	6.0	7.5	7.2
Total solids	50	400	300
Conductivity (S/cm)	45	700	600
Chloride	10	50	25
Alkalinity (total)	20	175	110
Hardness (total)	10	200	200
Colour (°H)	70	40	<5
Turbidity (NTU)	5	50	<5
Amm.N	0.05	0.5	0.05
NO_3N	0.1	2.0	0.5
Dissolved oxygen (percent saturation)	100	75	2
Biochemical oxygen demand (BOD)	2	4	2
22 °C colonies/ml (bacteria)	100	30 000	10

Source: Tebbutt (1983).

of sewers to a condition where the discharge to rivers is acceptable; often solid waste arises from this purification and there have been recent debates about how to dispose of this 'sludge', created by the protection of river quality.

As well as purely chemical contaminants of freshwater ecosystems there are organic threats to life and health ranging from *diseases* arising from human and animal waste to sewer rat infestations. Table 8.3 lists not only those infections relating to waste processes but those which are associated with water. It is often said that the majority of the world's hospital patients are suffering from water-based disease, a huge cost to the global environment.

8.1.3 History of water pollution controls

As revealed by Section 8.3 it is not unreasonable for mankind to rely on natural *recycling processes* to rid us of certain discharges of waste from our own bodies and our productive agricultural and industrial processes. Rivers exhibit a range of natural purification mechanisms (Fig. 8.2(a)) which operate on the wide range of contaminant inputs, transforming some permanently and storing others temporarily. Characteristic contaminants and pollutants have their own time-scales of impact as a result of these processes (Fig. 8.2(b)). The widespread reliance on river processes can be achieved without pollution if *loadings* are matched to the *capacity of the medium concerned* and the recycling/purification processes involved. Neither primitive nor modern man, however, has been able to use a fully rational matching of 'the need to contaminate' against 'the capacity to avoid pollution' in fresh waters.

From Roman times onwards the success of water-distribution engineering for irrigated agriculture in the prehistoric 'hydraulic civilizations' was translated into a system of sewerage to move human waste from the domestic source of its production to a convenient low point (so as to utilize a flow of water)—normally a river. Roman sewerage systems are even in place in such far-flung parts of their empire as the forts along Hadrian's Wall, Northumberland. In Britain, however, sanitary systems declined after the Romans left, mainly to an arrangement known as the earth-closet, effectively sewage disposal to land, though streams were still polluted with washings from the home, by flocks of animals, and by rural industries such as dyeing (see the records of convictions under by-laws listed by Parker, 1976).

Table 8.2(a). Some sources of river contamination and pollution

Origin of effluent	Type of effect			
	Chemical	Physical	Physiological	Biological
Sewage	Organic matter	Suspended solids Turbidity	Smell	Bacteria Viruses
Coal-mining	Dissolved inorganic matter	Suspended solids Bed deposits	Taste (iron)	
Food processing	Proteins Sugars Carbohydrates	Temperature		Bacteria
Metal industry	Metals Cyanide Phenols Ammonia	Colour	Taste (iron) Toxicity (cyanide)	
Organic chemical industry	Chlorinated organic compounds	Foaming	Taste (phenols)	
Electricity industry		Waste heat		

Source: Farrimond (1980).

Table 8.2(b). Discharges of trade effluent to sewers

Type of effluent	Total volume (Ml d^{-1})	Percentage of sewage works receiving this type of effluent
Food processing	166	56
Plating and metal finishing	195	53
General engineering	143	45
Laundering	25	40
General	72	39
Chemical and allied industries	184	31
Textiles:		
wool	72	20
cotton and man-made fibres	101	20
Brewing	111	19
Leather tanning and fellmongering	20	19
Paper and board	53	13
Printing and ink manufacture	26	8
Plastics	25	8
Iron and steel industries	25	5

Source: Farrimond (1980).

It is instructive to remain with Britain since, by experiencing the first Industrial Revolution, this nation developed certain basic principles of pollution management at an early stage, first for air and then for water. The Industrial Revolution led not only to a burgeoning scale and scope for industrial processes but the end of a relatively low density of population in both cities and countryside. Rapid

Plate 8.1. Outfall of a sewage treatment works (for a small town) into a tributary of the River Trent.

Table 8.3. Principal infectious diseases in relation to water supply

Disease	Frequency	Severity	Persons infected (millions)	Controllable with water and sanitation improvements (percentage)
Water-borne diseases:				
Cholera	L	H	n.a.	90
Diarrhoea	H	H	500	50
Typhoid	M	H	n.a.	80
Water-based diseases:				
Guinea worm	M	M	n.a.	100
Schistosomiasis	M	M	200	60
Vector-borne, water-related diseases:				
Malaria	H	H	300	n.a.
Filariasis	M	M	250	n.a.
Onchoceriasis	M	M	30	20
Trypanosomiasis	L	L	n.a.	80
Water-hygiene diseases:				
Roundworm	H	L	650	40
Hookworm	H	L	450	n.a.

Key: H = high; M = medium; L = low.

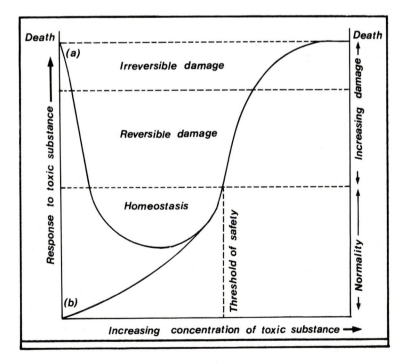

Fig. 8.2. The basic data of a toxicological approach to defining water pollution. Substance (*a*) is essential to well-being in small quantities (death results from its absence). Substance (*b*) is inessential to life.

urbanization and the rise of a poorly housed proletariat made the earth-closet inappropriate although night-soil collection from cities to disposal sites in the countryside persisted beyond the Victorian era.

To understand how the Victorians introduced the 'sanitary age' one must mention two other pressing urban problems: fire and drainage. Large sprawling estates of poor housing and dense business centres are prone to fire; public water supplies were first introduced in many areas for fire-fighting, with drinking and washing water fetched from the well or river. Urban streets became filthy, muddy, and rutted and hence getting rid of filth and getting rid of drainage water became one solution for two problems—a network of sewers draining to the nearest river. The supply of water to each house allowed a flushing mechanism for toilets, and factories also joined up to clean their premises and dispose of waste down the sewer.

A record of legislative development for Britain (Table 8.4) therefore records a fairly standard path for the rise of water-pollution risk as part of economic development. Religious and cultural differences introduced variants in the developing world but invariably disposal of waste to rivers eventually outstrips the purifying powers of the freshwater

medium, disease and death occurs, and remedies are sought.

The basic remedy, again largely derived from the 'sanitary age' of Victorian Britain, is the sewage-treatment works (Section 8.5.1). These harness the power of *natural purification processes* in an efficient system but they cannot deal either with the high inflow loadings produced after heavy rain (remember that surface-water drainage still 'powers' the flow of sewage to treatment works) or with the wide range of new and potentially toxic chemicals devised by industry every year.

Nation states and blocks such as the European Community vary in their attitude to legislative control over the inevitable water pollution resulting from the failure of sewage treatment to be thorough, comprehensive, and complete. Either limits are set on the concentrations of chemicals released by discharges into fresh water (or the sea) or there prevails a much less rigid approach (traditional in the UK) in which discharging is permitted in accordance with the perception of the ability of the receiving environment to 'cope'. Obvious cases of the need for extra care involve discharges to waters which are later used for drinking and to waters important for fisheries. Indeed, in the UK property

Table 8.4. A summary of water quality legislation in the UK, 1876–1989

Date	Statute	Date	Statute
1876	Rivers Pollution Prevention Act	1971	The Rural Supplies and Sewerage Act
1893	Rivers Pollution Prevention Act	1971	The Water Resources Act
1923	The Salmon and Fresh Water Fisheries Act	1972	The Clyde River Purification Board Act
1934	The Water Supplies (Exceptional Shortage Orders) Act	1973	The Water Act
1936	The Public Health Act	1974	The Control of Pollution Act (COPA)
1937	The Public Health (Drainage of Trade Premises) Act	1974	The Local Government (Scotland) Act
1944	The Rural Supplies and Sewerage Act	1975	The Reservoirs Act
1945	The Water Act	1976	The Drought Act
1946	The Water (Scotland) Act	1976	The Water Charges Act
1948	The River Boards Act	1980	The Water (Scotland) Act
1948	The Water Act	1981	The Water Act
1951	The Border Rivers (Prevention of Pollution) Act	1983	The Water Act 1987 The Control of Pollution (Landed Ships Waste) Regulations
1951	The Rivers (Prevention of Pollution) Act		
1951	The Rivers (Prevention of Pollution) (Scotland) Act	1985	The Food and Environmental Protection Act
1955	The Rating and Valuation (Miscellaneous Provisions) Act	1987	The Control of Pollution (Anti-Fouling Paints and Treatments) Regulations
1955	The Rural Supplies and Sewerage Act		
1960	The Clean Rivers (Estuarine and Tidal Waters) Act	1987	The Control of Pollution (Exemption of Certain Discharges from Control) (Variation) Order
1960	The Tidal Waters Orders		
1961	The Public Health Act	1988	The Control of Pollution Act 1974 (Commencement No. 19) Order
1961	The Rivers (Prevention of Pollution) Act		
1963	The Water Resources Act	1988	The Control of Pollution (Special Waste) (Amendment) Regulation
1965	The Rivers (Prevention of Pollution) (Scotland) Act		
1967	The Water (Scotland) Act	1988	The Public Utility Transfers and Water Charges Act
1968	The Sewerage (Scotland) Act	1988	The Transfrontier Shipment of Hazardous Waste Regulations
1968	The Tidal Waters Orders		
1968	The Water Resources Act	1989	The Water Act
1970	The Rural Supplies and Sewerage (Scotland) Act		

Source: River Pollution Control Unit, University of Stirling.

owners affected by pollution from upstream which endangers their livelihood can employ *riparian rights* in common law to bring a prosecution.

The most recent developments in freshwater pollution have been as follows:

- Increasing perception of the danger of *pollution of groundwater*.
- Increasing number of incidents of pollution from *agriculture* as opposed to industry.
- Increasingly holistic outlooks on the *relationship between river and marine pollution* (e.g. not diverting river outfalls to the nearest estuary or coast to claim a 'clean-up').
- *Integrated pollution control*—not permitting an emission from a plant to be burned in air as opposed to dumped in water or on land.

8.2 Toxicology

As the sensitivities of modern society to pollution of the environment become heightened by the publicity given to popular concerns there is danger of a situation of 'chemophobia' developing. Colleagues and students offered a glass of 'coloured chemicals' (in fact a fruit juice!) to drink from a laboratory beaker refuse point-blank. Since 'pure' water contains dissolved salts there is obviously a need for a scientific standpoint on chemical dangers—one provided by the field of toxicology.

8.2.1 Principles and methods of toxicology

The founding fathers of public health and sanitation knew nothing of toxicological testing; the water-borne infectious diseases which formed their main preoccupation were only known to be so because of epidemiological investigations. The most famous example is that of John Snow in London's Soho district in the cholera epidemic of 1854; Snow investigated a group of cholera cases whose water supply was a single pump in Broad Street. Only two cases, some miles away in Hampstead, appeared anomalous until Snow discovered that the sufferers were ex-inhabitants of Soho who liked the Broad Street pump water so much that they had bottles of it sent to them after they moved out of Soho!

Epidemiology, or rather *in vivo* (live) toxicological testing still forms a necessarily major part of our approach to water pollution; the first sign of a pollution incident may well be a large fish-kill (Plate 8.2) or a group of consumers complaining of unpleasant symptoms. The *in vitro* (laboratory) tests available to us are coarse and, many consider, cruel

Plate 8.2. The victims of a pollution-related 'fish kill' in the River Tyne are removed for examination. A successful prosecution was brought against the polluters.

to the test organisms whose lives are taken or stressed in order to measure the dose of chemical which should be considered unsafe (Fig. 8.3). There are too many naturally occurring and man-made chemicals (Table 8.5(a)) to permit all-pervasive toxicological tests but those which are tested can then be managed in a rational way according to a variety of criteria for expressing their harmfulness (Table 8.5(b)). The range of concentrations which become proscribed as damaging as the result of toxicological tests is very large (Table 8.5(c)); we often have difficulty in measuring the low concentration involved. Further doubt about testing comes from the differences in effectiveness between species (Table 8.5(d)), making the use of indicator organisms liable to error as well as cruelty. Clearly, too, an *ecotoxicological* approach is needed because a damaged organism is part of an ecosystem.

Many water-purity managers prefer to pass the final supply for human populations through tanks of live fish to act as continuous monitors of new or undetected toxic materials.

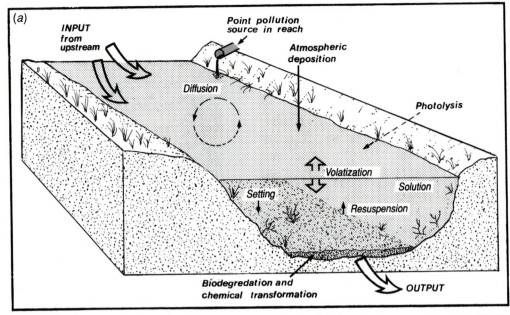

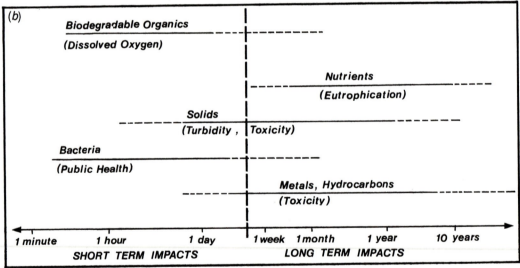

Fig. 8.3. Processes of (a) natural amelioration of pollution; and (b) time-scales of pollution impacts.

8.3 Pollution patterns and controls

The collection of waste in sewerage systems, its treatment and discharge to a river, lake or coastal sea, represents the traditional geographical pattern around which water-pollution control is (or is not?) managed. However, continued economic development and technological innovation present both new crises and opportunities for such point sources of pollution and a new set of geographical patterns concerning non-point or diffuse pollution, e.g. from rainfall, land-use, and other well-spread sources.

8.3.1 Point versus diffuse pollution

For the moment pathways and targets are taken for granted in order to indicate the basic difference in pattern of *point* and *diffuse* pollution. It should be

Table 8.5(a). Toxic substances present in industrial effluents

Substance	Source
Acids	Chemical industries, battery manufacture, mine-waters, iron and copper pickling wastes, brewing, textiles, insecticide manufacture
Alkalis	Kiering of cotton and straw, cotton mercerizing, wool scouring, laundries
Ammonia	Gas and coke production, chemical industries
Arsenic	Phosphate and fertilizer manufacture, sheep dipping
Cadmium	Metal plating, phosphate fertilizers
Chlorine (free)	Paper mills, textile bleaching, laundries
Chromium	Metal plating, chrome tanning, anodizing, rubber manufacture
Copper	Plating, pickling, textile (rayon) manufacture
Cyanide	Iron and steel manufacture, gas production, plating, case-hardening, non-ferrous metal production, metal cleaning
Fluoride	Phosphate fertilizer production, flue-gas scrubbing, glass etching
Formaldehyde	Synthetic resin manufacture, antibiotic manufacture
Lead	Paint manufacture, battery manufacture
Nickel	Metal plating, iron and steel manufacture
Oils	Petroleum refining, organic-chemical manufacture, rubber manufacture, engineering works, textiles
Phenols	Gas and coke production, synthetic resin manufacture, petrolelum refining, tar distillation, chemical industries, textiles, tanning, iron and steel, glass manufacture, fossil-fuel electricity generation, rubber processing
Sulphides	Leather tanning and finishing, rubber processing, gas production, rayon manufacture, dyeing
Sulphites	Pulp processing and paper mills, viscose film manufacture
Zinc	Galvanizing, plating, rubber processing, rayon manufacture, iron and steel production.

Source: Hellawell (1986).

Table 8.5(b). Terms used in expressing toxicity data

Term	Definition	Abbreviation
Acute (short-term) tests		
Median effective concentration*	Concentration at which a specified effect is observed in half the population within a given time	48 hr EC_{50}
Median lethal concentration*	Concentration at which half the population will die in a specified time	96 hr LC_{50}
Median tolerance limit	Concentration at which half the population will die in a specified time	96 hr TL_m
Median effective time	Time taken for an observed effect to occur in half the ET_{50} test population at a given concentration	(TL_{50})
Median survival time, or median period of survival	Time taken for half of test organisms to die at a given concentration	LT_{50}
Chronic (long-term) tests		
Incipient lethal level	Concentration of toxic substance or other potentially lethal condition (e.g. low oxygen) which organisms could tolerate indefinitely	ILL
Maximum acceptable (allowable) toxicant concentration	(Maximum) concentration of the toxic substance which is acceptable within the environment in order to ensure no harm to organisms	MATC

* Sometimes called dose, in which case LD, ED etc.

Source: Hellawell (1986).

Table 8.5(c). Consent limits formerly applied to effluents into sewers and rivers by Thames Water, UK

Parameter	To sewers	To rivers
Effluents (mg/l⁻¹)		
Sediment	200–1000	5–30
Cyanide	10	0–1
Free ammonia	50–100	0.1–10
Available chlorine	50	0–0.5
Formaldehyde	200	nil
Grease/oil	50–400	0–2
Chromium	1	0.1–<1
Lead	10	3
pH	6–11	6–9
Temperature (°C)	<43	<31

Table 8.5(d). Species differences in toxicity of ipomeanol

	LD_{50}mg kg^{-1}*	Location of tissue damage		
		Liver	Kidney	Lung
Rabbit (New Zealand White)	40	—	—	•
Mouse (A/J strain)	20	—	•	•
Rat (Fisher strain)	12	—	—	•
Hamster (Syrian Golden)	140	•	—	•
Guinea pig (Hartley)	30	—	—	•

* The ipomeanol was administered intraperitoneally in 25% aqueous propylene glycol to all species.

Source: Timbrell (1989).

emphasized that although lakes are not shown they are susceptible to pollution from the same kinds of source. In Britain, however, our major lake systems tend to be in remoter, rural areas, away from the major point sources shown but vulnerable to a number of diffuse sources of pollution, including recreational use of the lakes themselves. To the inhabitants of, for example, the cities around North America's Great Lakes, however, lake pollution is a major environmental problem. We should also emphasize, following the broad habitat coverage of Chapter 7, that wetlands are vulnerable to all the major forms of pollution and may well be a favoured alternative for discharges where they cannot be reclaimed for intensive uses of land.

The major differences in the properties of (and therefore control strategies for) point- and diffuse-source pollutants are shown in Table 8.6. Point sources are readily identifiable compared with diffuse sources, making control easier in a perfectly regulated society. The truth is less efficient, especially where the point source is an ancient sewer to which unknown numbers of unofficial connections have been made over the years. Whilst the pollutants emitted by point- and diffuse-sources are generally of a different nature (respectively mainly faecal matter, metals, etc., and nutrients, pesticides, etc.) the two sources are considered together in any comprehensive scheme of river management. Whilst the diurnal fluctuations of urban, point-source pollution can be controlled to fall below a certain level fixed by water-quality (purity) objectives, if diffuse-

Table 8.6. Point-source and diffuse-source pollution of freshwaters: key differences

Characteristic	Point source	Diffuse source
Types of waste	Domestic, urban, industrial	Agricultural or urban via-atmosphere: gaseous.
Concentration of toxic material	High	Lower but subject to progressive accumulation or high impact through regular exposure of target (e.g. drinking-water).
Pattern of discharge in time	Tendency to be episodic—during factory hours etc.	Tendency to be continuous or steady but with episodes resulting from e.g. weather sequences such as dry . . . wet.
Control opportunities	Legal control of identifiable, registered sources (or economic control through fines/pollution taxes)	Source identification problematic. Educate land users/mitigate by land treatments including economic incentives.

source pollution from upstream gradually raises the concentration of pollutants, the overall effect is to return the river to a failed, possibly illegal, state in relation to the objectives.

8.3.2 Monitoring freshwater pollution

It is essential to define monitoring and, as suggested by Hellawell (1991) it is useful to distinguish it from survey:

> *Monitoring*: Intermittent (regular or irregular) surveillance carried out in order to ascertain the extent of compliance with a predetermined standard or the degree of deviation from an expected norm.
>
> *Survey*: An exercise in which a set of qualitative or quantitative observations are made, usually by means of a standardized procedure and within a restricted period of time, but without any preconception of what the findings ought to be.

Monitoring freshwater pollution requires a thorough appreciation of the sources, pathways, and targets for a myriad of constituent pollutants. Table 8.2(a) indicates a restricted list of those contaminants which might cause pollution if allowed to reach concentrations above those suggested by toxicological tests. Section 8.2 has already pointed out the value of an all-embracing technique such as *in vivo* use of live trout in water-treatment works.

Because monitoring is purposeful, the selection of a monitoring *technique, location*, and *frequency* usually depends upon pollution-control legislation. If emission controls are imposed, e.g. on sewage treatment works, 'end of pipe' *monitoring of discharges* is appropriate. However, if limited pollution of pathways such as the river or lake itself is the basis of control it is essential to *monitor the ambient, or background quality* of the environment into which a discharge is made and the impact of the discharge once it has been added.

Monitoring targets is the least well developed scientifically; targets should be protected by general toxicological research and by source/pathway monitoring. That this protection is unreliable may be judged from the fact that the freshwater environment is still losing biota from both episodic kills (Plate 8.2) and long-term extinction and the persistence of confirmed or alleged water-borne or water-related illness in human beings. Targets are currently more the focus of *survey* than of *monitoring*; human targets may actually participate in field experiments as when bathers are asked to record the ill-effects of immersion in coastal waters. Just as

John Snow was able to determine the source of cholera in Soho from disease records (see Section 8.2) modern concerns for pollution to be confirmed as a source of water morbidity/mortality, e.g. childhood leukaemia and Alzheimer's Disease, have suggested that properly compiled geographically indexed medical records are essential.

Monitoring freshwater pollution also faces a number of purely practical scientific and operational problems such as:

(1) The availability of robust and repeatable analytical tests.
(2) Instrumentation at an affordable price to perform them.
(3) Protection for instrumentation against environmental extremes and vandalism.
(4) Spatial variability, over short distances, of the parameter being monitored.
(5) Temporal variability of the parameter being monitored.
(6) Acceptance by the legal processes of pollution regulation of the evidence from a monitoring system, especially of the degree of severity of pollution and its frequency of occurrence.

Numbers (2) and (3) on this list are nowadays quasi-commercial items: if the need is sufficient and 'the price is right'—'we have the technology'. Items (1) and (6) are beyond the scope of this book but severely constrain the whole pollution-control process (see Hawkins, 1984). *Regulatory action* operates in a socio-political context and differs fundamentally from *crime control*. As a simplistic example, consider the difference in your own perception of a breath test after driving erratically and an arrest for breaking a jeweller's window (even if theft is not intended)!

Items (4) and (5) are central concerns for those convinced of the relevance of a geographical context to pollution control. Fig. 8.4 illustrates the problems involved by combining the layout of a contaminated river, the spatial location of a sampling station, and the likely variability of dissolved constituents at key points, in the channel cross-section and through time. Whilst we have stressed monitoring here as a contemporary, pro-active, and dynamic activity several recent pollution issues have raised the question of *retrospective monitoring*. The first issue with which environmental reconstruction became familiar was the impact of terrestrial pesticides on birds through their breeding success; egg collections dated to specific eras of history were used to indicate the process of shell-

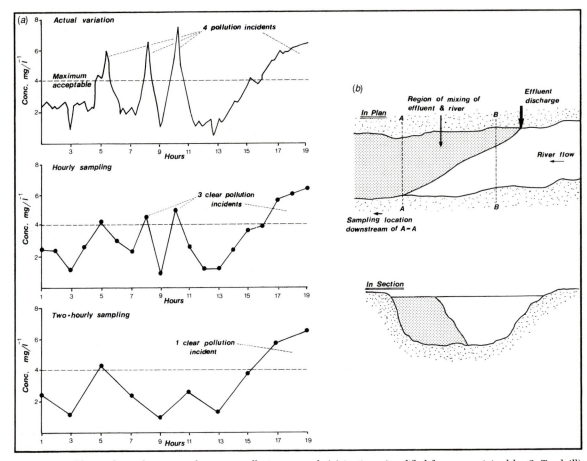

Fig. 8.4. Problems of sampling rivers for water pollution control, (a) in time, (modified from an original by S. Trudgill) and (b) in space (after D.o.E., NWC, 1980).

thinning which coincided and accelerated with the introduction and use of harmful agrochemicals.

Freshwater examples have occurred in connection with acidification. The work of Battarbee (see Renberg and Battarbee, 1990) on diatom records from lake sediments has confirmed acidification as a gradual post-glacial phenomenon, through weathering of bases, accelerated in the 1850s as a result of deposited acidity from the industrial sources developed at that time. Once again ornithological records have also been of interest—the decline of the dipper (Ormerod and Tyler, 1986) mirroring that of pH in a Welsh stream—see Fig. 8.5.

Use of surveys as well as of monitoring in the control of pollution is indicated by the quinquennial data collected for most European rivers such that river-reach lengths can be categorized according to classes representing their chemical or biological quality. In England and Wales such surveys date

back to 1958, establishing a good basis for assessing the efficacy of clean up campaigns and the strictures of the law (Table 8.7(a) and (b)). Nevertheless, such extensive surveys are bound to be based on far fewer parameters than would be utilized to monitor a discharge from a factory and the methods used to calculate which lengths of river should be accorded to the class suggested by the sampled values also lead to controversy. For the first time the 1991 survey of rivers in England and Wales has raised biological parameters to a status equalling that of chemical determinants.

8.3.3 Pollution control: contrasts in international practice

It is often claimed that primitive societies have the advantages in pollution control of being thoroughly and obviously integrated with their natural environment. This integration gives them a cultural basis

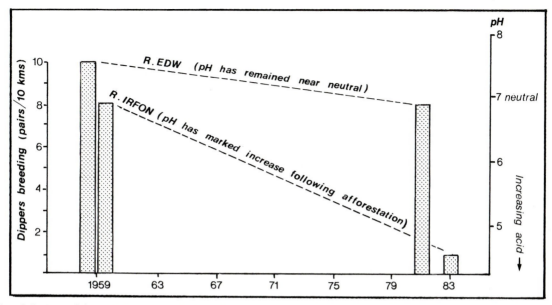

Fig. 8.5. Use of historical, indirect biological data to indicate pollution trends: the case of dippers and stream acidification (after Ormerod and Tyler, 1989).

upon which to set *norms* and develop *practices*, many of which are now under the scrutiny of anthropologists.

The development process both creates the waste which causes pollution and distances society from its major impacts. Environmentalists have in recent years called for the developed world to debate and adopt a new ethic which would replace that lost in our departure from the 'primitive'. Such an ethic would, because the environment is now a global issue, need to be international. Meanwhile the cultural characteristics and political systems of different nation states tend to erect different standard practices towards the environment, including pollution control.

In terms of freshwater pollution such contrasts are often revealed by the media in terms of practices in the United Kingdom and the European Communities (Haigh, 1990; Newson, 1991a). The UK is conventionally described by pressure groups as 'The Dirty Man of Europe' as a result of certain perceptions of the contrast with other nations in the matter of control practices.

The UK's approach to pollution control developed

Table 8.7(a). Water quality in England and Wales, 1958–1980 surveys using former classifications

Class	1958		1970		1975		1980	
	km	%	km	%	km	%	km	%
Non-Tidal Rivers and Canals								
Unpolluted	24 950	72	28 500	74	28 810	75	28 810	75
Doubtful	5 220	15	6 270	17	6 730	17	7 110	18
Poor	2 270	7	1 940	5	1 770	5	2 000	5
Grossly polluted	2 250	6	1 700	4	1 270	3	8 10	2
TOTAL	34 690	100	38 400	100	38 590	100	38 740	100
Tidal Rivers								
Unpolluted	1 160	41	1 380	48	1 360	48	1 410	50
Doubtful	940	32	680	23	780	27	950	34
Poor	400	14	490	17	420	15	220	8
Grossly polluted	360	13	340	12	280	10	220	8
TOTAL	2 850	100	2 880	100	2 850	100	2 800	100

Table 8.7(b). Water quality in England and Wales, 1980–1990 surveys using new classifications

Class	1980 km	1980 %	1985 km	1985 %	1990 km	1990 %
Freshwater Rivers and Canals						
Good 1A	13 830	34	13 470	33	12 408	29
Good 1B	14 220	35	13 990	34	14 536	34
Fair 2	8 670	21	9 730	24	10 750	25
Poor 3	3 260	8	3 560	9	4 022	9
Bad 4	640	2	650	2	662	2
TOTALS	40 630	100	41 390	100	42 434	100
Estuaries						
Good A	1 870	68	1 860	68	1 805	66
Fair B	620	23	650	24	655	24
Poor C	140	5	130	5	178	7
Bad D	110	4	90	3	84	3
TOTALS	2 730	100	2 730	100	2 722	100

Note: Lengths are rounded to the nearest 10 km and may not sum to totals.

as a reaction to the extremes of degradation of local environments during the Industrial Revolution, particularly as the result of air pollution (see Brimblecombe, 1988). There were four main principles:

- No action until proper scientific proof
- Polluter pays for remedial action
- Control as near to source as possible
- Use of 'best practicable means' of control.

When the European Commuunities first formulated a strong environmental policy (after the Stockholm Conference of 1972) its principles were said to be:

- Control at source, i.e. of emissions
- Polluter pays
- Precautionary action in advance of full scientific evidence
- No production advantage by member states at the expense of the environment.

Contrasts in the two perspectives arise largely as the result of differing emphasis on the *burden of scientific proof* and differing anticipations of the *impacts of control on polluters*. The outcome is that the EC tends to use what has become known as *the Precautionary Principle*, a derivation from German public policy (see Royal Commission on Environmental Pollution, 1988) whereas the UK persists with variants of 'best practicable means'; the former is strict and applies emission standards, the latter is seen as lax and relates desirable ambient standards to achievable control technologies and policing of legislation.

Before briefly cataloguing some typical water pol-

lution control issues which have highlighted this contrast of approaches it is worth while to stress a geographical difference between the UK and mainland Europe: the former is part of a group of islands with short, steep rivers offering good transport and dilution potential for pollutants whilst the latter is part of a very large continental mass with numerous lengthy rivers, crossing international borders (or forming them).

8.3.4 Towards emission standards? Contrasts in international regulation

Four categories of aim for water-pollution control strategies are laid out in Box 8.1; *criteria* are common to the remaining three categories which are tabulated in order of increasing stringency.

Box 8.1. *Categories of aim for water-pollution control strategies*

1. *Criteria*: scientific data evaluated to derive the recommended limits for water uses.

2. *Water-quality guideline*: numerical concentration or narrative statement recommended to support and maintain a designated water use.

3. *Water-quality objective*: a numerical concentration or narrative statement which has been established to support and protect the designated uses of water at a specified site.

4. *Water-quality standard*: an objective that is recognized in enforceable environmental control laws of a level of government.

However, whether relatively easy *guidelines* or relatively tough *standards* are being followed there is always a further categorization depending on the *designated uses* for the river or lake; Box 8.2 illustrates how the upper limits of cadmium concentration which might be set as a standard would vary according to the most sensitive user category in the list. It is this philosophy of controlling pollution with respect to targets set by ambient conditions which distinguishes the UK approach to water-pollution control. The EC approach is based on standards set for pollutant *emissions* (everywhere) rather than for the water body into which the pollutants are discharged.

exceeded (known as MACs—Maximum Allowable Concentrations), are now set by European controls, e.g. for discharges from sewers.

Table 8.9 shows how, for a low-cost infrequent sampling routine (ten samples per year—not uncommon in the UK), a considerable number of failures on individual parameters can give a credible performance unless the summation of pass/fail results is used, pulling compliance down to an illegal 30 per cent. It is essential, therefore, when environmental groups or the general public are consulting pollution-control performance data that careful attention is paid to the definition of success and to the basic sampling design.

Box 8.2. *Cadmium concentration in water required for different uses, as recommended by the US National Academies of Sciences and of Engineering to indicate the most demanding uses*

Water use	Maximum concentration (g/l)
Livestock water supply	50
Human potable supply	10
Freshwater aquatic life (in hard water, > 100 mg/l as $CaCO_3$)	30
Freshwater aquatic life (in soft water, < 100 mg/l as $CaCO_3$)	4
Crustacea and eggs and larvae of salmonids (in hard water)	3
Crustacea and eggs and larvae of salmonids (in soft water)	0.4
Marine aquatic life required for human consumption	0.2
Marine aquatic life required for human consumption, with concentrations of zinc or copper exceeding 1 mg/l	0.02

Source: Water Research Centre.

Once the principles of a pollution control system are decided the regulators must set critical values which prompt emergency or punitive action. As an example of a relatively stringent set of criteria and their critical values, Table 8.8 lists the American and European freshwater fisheries guidelines. Nevertheless, despite the similarity of fish metabolism on both sides of the Atlantic there are clear differences in both the scope and values of the listed criteria; European standards are generally considered very fastidious by international comparison.

It will be noted, too, from the table that certain of the American values are keyed as '95-percentile values', meaning that they must be met for more than 95 per cent of all occasions when a sample is analysed. However, it has become obvious that on the 5 per cent of occasions when pollution is permitted by this loophole, serious biological damage can be done and so absolute values, never to be

There have been battles, therefore, between European stringency and British pragmatism, not only over whether emissions should be uniformly restricted but over the assessment of compliance. The UK has compromised recently by setting *emission standards* for seriously damaging pollutants and strengthening river water-quality objectives by making them legally enforcible rather than a 'wish list' for improvement. The British talent for compromise has now emerged as enforceable objectives for most parameters, except emission controls on highly toxic substances.

8.4 Examples of point pollution of freshwater systems

We now turn to a more detailed examination of the pollution danger from typical sources. In all cases it

Table 8.8. Water-quality criteria for freshwater fish and aquatic life based on EPA (USA) and EIFAC (Europe) guidelines

Substance	Criteria for protection of fish	
	EIFAC	EPA
Aldrin Dieldrin		0.003
Ammonia (un-ionized)	25	(20)
Arsenic		
Cadmium:		
hardwater	1.5*	1.2
softwater	0.9*	0.4
Chlordane		0.01
Chlorine	4	2
Chromium		(100)
Copper:		
hardwater	112*	0.1 96 hr LC_{50} (c.10)
softwater	22*	0.1 96 hr LC_{50} (c.2)
Cyanide		5
DDT		0.001
Endosulphan		0.003
Endrin		0.004
Heptachlor		0.001
Lead:		
hardwater		0.01 96 hr LC_{50}
(c.500)		
softwater		0.01 96 hr LC_{50} (c.5)
Lindane (HCH)		0.01
Malathion		0.1
Mercury		0.05
Methoxychlor		0.03
Nickel		0.01 96 hr LC_{50}
(c.100)		
Oil		0.01 96 hr LC_{50}
(c.100)		
Parathion		0.04
PCBs		0.001
Phenols	1000	1†
Phthalate esters		3
Selenium		0.01 96 hr LC_{50} (c.20)
Silver		0.01 96 hr LC_{50} (c.0.1)
Sulphide		
(undissociated H2S)		2
Zinc:		
hardwater	50*	0.01 96 hr LC_{50} (c.25)
softwater	20*	0.01 96 hr LC_{50} (c.3)

* percentile values.
† Value selected to avoid tainting of flesh.
Note: Values are in g litre^{-1}. Where application factors of the 96 hr LC_{50} are specified approximate values are given in brackets.
Source: Hellawell (1986).

is important when considering control or reduction strategies to remember the importance of integrated pollution control. Thus a reduction in the amount of sewage sludge or cattle slurry entering water may mean an equally damaging discharge of that material to land or, via combustion in waste incinerators, to air.

8.4.1 Human cycling: water purification and sewage treatment

Nations vary in the proportion of water that they draw from surface and underground sources (see Section 3.3). The use of groundwater has generally implied that the benefits of natural filtration are gained at the loss of energy required for pumping from wells and boreholes. Surface waters are generally cheaper to supply but inevitably require purification: 'shocking' statistics abound on the number of human beings through which river water has passed en route for the sea—six for the Thames, thirteen for the Rhine! *Water purification* and *sewage treatment* are inextricably linked in most densely-populated, developed nations; often they are administered by the same organization and involve certain common processes (Fig. 8.6).

Fig. 8.6 shows a simplified diagram of both operations, assuming surface sources and discharges. It will be noted that water purification often adds more chemicals than it subtracts; the 'raw' water of the surface source is maintained in an unpolluted (though possibly contaminated) state by controls on river pollution and so useful chemicals are added during purification to aid *clarity* (e.g. aluminium sulphate), *neutrality* (e.g. lime), and *dental health* (fluorine). In addition the water may be *disinfected* by use of gases such as chlorine (to remove pathogenic organisms).

These chemical processes are carefully controlled and are applied after two simple but very effective physical processes—*storage and filtration*; storage not only begins the process of clarification but is fatal to many pathogenic organisms which rely on oxygenation to prosper in the river itself. Recent concerns have derived from general 'chemophobia' but specifically about the possible relationship between the aluminium ion (not completely removed after clarification) and Alzheimer's disease, and certain

Table 8.9. Parameter or sample compliance

	Ten sample results	Compliance (%)
BOD	— — — F — — — F — —	80
SS	— — F — F F F —	50
Ammonia	— F — — — F — — — —	80
Sample	— F F F — F F F F —	30

Notes: F = fail.   BOD = Biochemical Oxygen Demand
SS = suspended solids
To get over these problems, 'look-up' tables have been created which allow for sampling error, and all consents are judged in terms of 95 percentiles for each determined separately. So, for example, if only 4–7 samples are taken in a year, 1 is allowed to fail for each determinand, which could mean that *no* sample fully complies!
Source: Rees (1989).

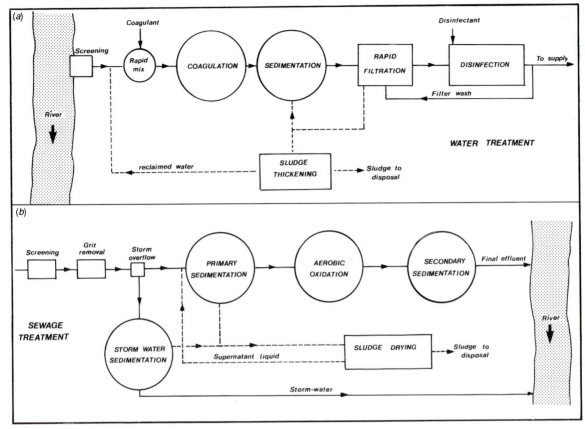

Fig. 8.6. Systematic presentation of major processes (a) in a water-treatment works, and (b) in a sewage-treatment works. Some of the processes are identical.

mutogenic (carcinogenic) chemicals produced from highly organic waters after chlorination (see Jones, 1991).

We have also referred to the use of *live trout* as arbiters of the quality of water entering supply; in the UK this system has recently been quantified by systems which monitor the metabolism of the test fish, rather than using the observations of their health by workers in the treatment plant.

It should be noted, in relation to integrated pollution control, that water purification itself yields a potential pollutant: *sludge* from the clarification and settling processes (Plate 8.3); this can be disposed of to landfill but of course is cumulative and has a high metals content.

Sewage treatment yields much larger quantities of sludge; the constitution of sewer discharges is shown in Table 8.10. The solids comprise not only faecal matter, paper, sanitary towels, and condoms from lavatories but, in combined 'storm' sewers, the

Plate 8.3. Sludge removed from river water ('raw' water) as part of purification for domestic supply.

Table 8.10. Contents of typical sewer discharge to river and comparative figures for crude, untreated sewage

Parameter (mg/l^{-1}) sewage	Effluent	Crude
Biochemical Oxygen Demand	20	300
Suspended solids	30	400
Ammonium N	5	40
Nitrate N	20	<1
Phosphate	6	10

Source: Tebbutt (1983).

litter from our streets (including dog faeces) and a considerable sediment load (see Ellis, 1991).

The sewage treatment works is essentially modular—three main components may be applied to the purification problem: primary, secondary, and ter-

Table 8.11(a). Sewage treatment: modular stages and component processes

Processes	Outputs	Utilization
Primary		
Screening	Materials	Landfill
Grit removal	Grit	
Sedimentation	Sludge	
Secondary		
Lagoons		Sea dumping (now proscribed)
Biochemical treatment	Sludge	Spread to land (free to farmers)
Percolating filter		Incinerated
Tertiary		
Biological denitrification	Sludge	Incinerated Spread to land
Coagulation/ sedimentation		
Microstraining	Recovered water	Can be purified for re-use
Disinfectant		

tiary (Table 8.11(a)), forming an incremental series with primary treatment providing the least satisfactory cleansing and tertiary treatment being suitable for the most sensitive locations (see Table 8.11(b)).

The criteria by which discharges to rivers from sewage treatment works are judged have, until recently, been based upon work done in the nineteenth century in the UK. As a prelude to the Royal Commission on Sewage Disposal's recommendations for discharges (1912) it was assumed that there would be an eightfold dilution by the river concerned (every river, at all times)—an enormous simplification of the case in practice.

The advantage of the simplicity of these criteria was that they were operable at comparatively little cost and were legally 'transparent' to regulators and regulated. However, as industry began to add discharges to sewers and a 'cocktail' of chemicals began to enter storm drains from urban surfaces (consider the toxic freight which might spill on to roads) these criteria became inadequate, as did the sampling regime. The modern consent system by which UK sewage discharges are licensed now extends to dissolved oxygen as well as Biochemical Oxygen Demand (BOD), to ammoniacal nitrogen, pH, and suspended solids (SS).

With the stricter criteria and compliance calculations in place the geography of point-source river pollution should, in practice, lead to better planning of industrial location and urban development. Each point source produces a 'sag' in water quality (Fig. 8.7) as it enters the river but each can be calculated according to the diluting potential of the receiving water to ensure the viability of the river system under a variety of flow conditions.

There are wide national variations in the propor-

Table 8.11(b). Probable domestic sewage treatment for various receiving waters

Receiving water	Typical effluent standard		Probable treatment
	BOD	SS	
Open sea	—	—	Screening or comminution
Tidal estuary	150	150	Screening or comminution primary sedimentation, sludge disposal by dumping on land or at sea.
Lowland river	20	30	Screening or comminution primary sedimentation, aerobic biological oxidation, secondary sedimentation, sludge stabilization, sludge disposal on land, or at sea, or by incineration.
High-quality river	10	10	As for lowland river with addition of tertiary treatment by sand filtration, grass plot irrigation, or lagoons.

Source: Tebbutt (1983).

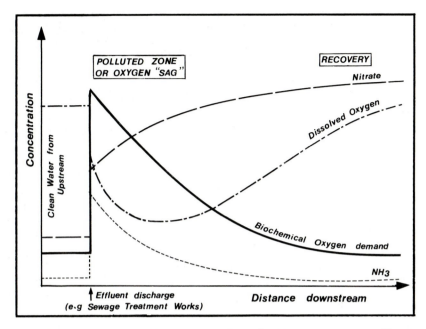

Fig. 8.7. The impact of a point-pollution source on the downstream properties of rivers (typical of the outfall from a sewer system).

tion of sewage which receives the three stages of treatment. In the UK almost 100 per cent receives some treatment and 80 per cent receives secondary treatment; in the Republic of Ireland the figures are 60 per cent and 5 per cent with 40 per cent going to land disposal direct. Even in the USA only three-quarters of the population is served by sewage treatment, whilst in Belgium only one third is served.

8.4.2 Industrial wastes

Industries slowly perceived the advantages of connecting to the sewerage system as the volume and variety of their wastes increased in the nineteenth century (in the case of the UK). The paradox is that sewage treatment did not itself experience a parallel increase in ability to remove industrial products. As described in Section 8.4.1, the conditions for treated sewage discharges to rivers have focused around only two sample parameters, oxygen and suspended sediment.

By contrast industrial additions to sewers (and deposits/spillages from roads and buildings via storm sewers) present a highly complex problem of purification. A major complication can occur because of the unwillingness of industrialists to supply details of a production process (many of the details of which can be inferred from the waste products), or pay for purification. In the case of major industrial complexes industry constructs and operates its own treatment works with the water-

pollution control authorities monitoring effluent therefrom (Plate 8.4).

8.4.3 Animal feeds and wastes: silage and slurry

In a traditional, extensive livestock-farming system there is an inbuilt balance between stocking density and the recycling of nutrients through urine and faeces to boost the fertility of herbage. Modern, intensive livestock farming differs, not only because the cycle is 'inflated' (more livestock, more herbage) but because mixed farming and disease control/avoidance of climatic stress permits indoor housing of stock for parts or all of the year.

The result is that some livestock farms have become unbalanced; whilst animal wastes are collected from housing lots and distributed back to land there is often insufficient land resource to take these applications (because it is in arable cultivation or may have nature-conservation designations forbidding the use of fertilisers) or there are problems of timing applications in a seasonal climate. The answer is storage of wastes but provision of such storage is expensive. As a result cheap *slurry lagoons* have been constructed which may leak or rupture, causing a severe pollution incident in any nearby watercourses. A similar storage problem exists with silage—the grass feed harvested without drying (as for hay) and therefore releasing liquors in storage as it decays.

The potential pollution effect from slurry (and

Plate 8.4. An industrial sewage treatment works.

silage liquor) may be judged from Fig. 8.8 in comparison with human wastes. An accidental spillage of such a deoxygenating liquid (which also has a highly acid reaction) clearly pollutes all watercourses, ponds, and lakes into which it passes. For this reason most nations with productive farm systems have introduced regulations which curtail the freedom of farmers to locate and build their own silage and slurry stores, which specify the construction standards and provision of emergency storage capacity, with bunds to contain spills, and which set minimum distances to the nearest freshwater body.

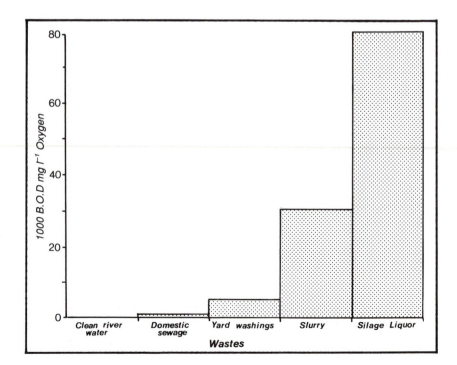

Fig. 8.8. The relative strengths of agricultural wastes, domestic sewage, and clean river water as measured by biochemical oxygen demand (BOD).

Such restrictions are also being imposed on a variety of other noxious chemicals used on farms (Conway and Pretty, 1991) including fuel oils, pesticides, and sheep-dip chemicals.

8.4.4 Land disposal of wastes: hydrological pathways

This section both completes our sample survey of point pollution sources and sets up coverage of diffuse sources. An alternative disposal mechanism for domestic and industrial sewage and for farm wastes is clearly the land itself; since many of the contents of these wastes are beneficial and because recycling is in the long-term interests of resource management the idea is an attractive one. However, revision of our knowledge of hydrological processes from Part I allows us to forewarn that land and water cannot be considered separately, especially as alternative destinations for pollutants from point sources. Whilst very careful technical attention is lavished on the design of sewage treatment works, landfill waste disposal is 'the Cinderella of the waste treatment industry' (Fleming, 1988)—taking an

'out of site, out of mind' approach to the inevitable relationship between landfill (*leachate* solutions, rich in contaminants (Fig. 8.9(a)) and surface/subsurface hydrological pathways.

It is common for new landfill sites to be given a hydrogeological appraisal and an impermeable lining where water supplies are threatened but, as Fleming stresses, the sheer scale of modern sites means that the hydrology of the site and its surroundings changes during the life of the scheme. Management systems with respect to hydrological protection (Fig. 8.9(b)) are also needed in *perpetuity*, and there are hopeful signs of developing beneficial products such as landfill as a result; clearly a much greater emphasis on recycling of wastes will help ease the problem. There is growing evidence, too, of groundwater pollution from accidental or deliberate use of holes in the ground for the disposal of wastes from medical, engineering, and domestic wastes; for example, chlorinated solvents are being found in concentrations likely to cause pollution in the Birmingham aquifer (Rivett *et al.*, 1990).

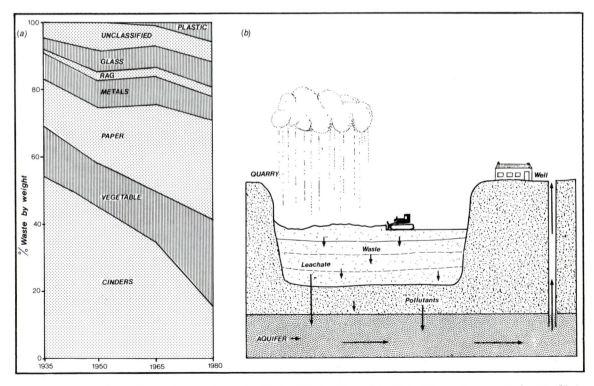

Fig. 8.9. Water pollution by solid waste disposed of by landfill. (a) The composition of domestic waste in the UK. (b) A cartoon cross-section of a landfill site.

8.5 Pollution from diffuse sources

Pollution from point sources may be chronic and highly visible; that from diffuse sources is seldom so, except where it has been progressive and where an environmental threshold has been crossed relatively suddenly—as in the case of an algal bloom in a eutrophying lake. Since it operates mainly through the media of land and air it is considered to be separated from the management of water resources, and inter-agency rivalry in its control is rife. The part played by the atmosphere also internationalizes the issue, as is clear from the case of acidification. Before commencing this section it might be advisable to consult Section 3.8 again on the role of hydrological pathways in determining chemical water quality.

8.5.1 Agriculture and forestry as sources: land use versus land management

Rural activities scarcely match well in our minds with stressed environments; both farming and forestry appear to be the desirable activities from which the majority of mankind in developed nations migrated to unhealthy cities to work in polluting industries. The risk of pollution from modern rural activities arises as a result of the intensification which has been necessary in production to support the large human numbers in those urban and industrial centres. It is significant that public perception has become much less sympathetic to farmers since the revelation of the political problems raised in Western countries by food surpluses; it is

possible that we pay four times for food—once to support its intensive production, once to buy it, once to pay for the pollution its production causes, and once to dispose of surpluses! Lack of public sympathy for plantation forestry may either involve its aesthetic effects on the open landscape (e.g. in the UK) or its inroads upon natural wilderness (e.g. in the natural forests of Scandinavia and North America).

When dealing with the environmental effects of rural activities we must distinguish between land *use* and land *management*; this distinction aids environmental control strategies (Newson, 1991). For example forestry is a land *use* but ploughing, draining, and roading are parts of forestry land *management*. As we have seen above, livestock farming is a land use but the generation of slurry disposal problems is a corollary of the type of livestock management practised.

Further distinctions are necessary when dealing with rural, diffuse sources of pollution between stages of the crop, husbandry, or forest cycle; Table 8.12 makes these clear.

8.5.2 Nitrate pollution

This is a diffuse-source pollution issue which has predominantly affected western European agriculture since World War II; the crisis proportions which it reached in the late 1980s were anticipated early in the post-war period by those who observed that agricultural recovery was largely supported by the use of artificial fertilizers. Another feature of post-war agricultural intensification—the use of pes-

Table 8.12. Production/management cycles and diffuse sources of water pollution in rural areas

Land-management stage	Farming		Forestry	
	Arable	Livestock	Plantation	Natural
Ground preparation	Sediment loss from ploughing	(Infrequent: less impact than arable)	Drainage/ditching	n.a.
Planting/ establishment	Sediment loss from autumn-sown crops	Cover crop often grown as pasture establishes	Danger of fertilizer/ pesticides applied to prevent competition	n.a.
Maturity	Applications of fertilizers/ pesticides	Soil erosion from heavily grazed root crops in winter	Indirect acidification from canopy scavenging	Increased concentrations of chemicals as evaporation increases
Harvesting	Release of nutrients from crop residues	Wilting agents sprayed for silage harvesting	Danger of soil damage, landslides, mineralization of nutrients and dispersion of fungicides.	
Between crops	(formerly air pollution from straw burning)	Plough up releases nutrients	(Reaction highly dependent on style of replanting, ground cover, species)	

ticides—is now alleged by many to be leading to future water-pollution crises.

The apparently slow buildup of the problem in raw waters (Fig. 8.10) and drinking-water supplies has much to do with the dominant pathway followed by the pollution itself in the lowland areas where arable crops thrive, i.e. via aquifer systems. Whilst nutrient chemicals reaching surface water systems may be 'stripped' by wetlands or utilized by stream biota, groundwater flow routes do not offer

these opportunities. Another reason for the sudden precipitation of the nitrate pollution control crisis was the growing but imprecise toxicological evidence for its effects on humans via drinking water, e.g. as a cause of blood disorders in babies and cancers in adults (the evidence linking nutrient pollution to biotic damage in freshwater environments is less contentious but equally imprecise—see Section 8.7).

The monitoring of nitrate levels in drinking-water

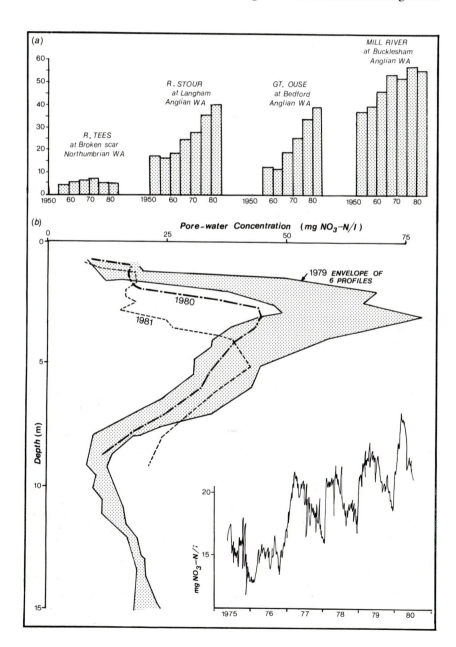

Fig. 8.10. The impact of nitrate pollution in the UK (a) on river water (after D.o.E., 1986) and (b) on groundwater. The steady downward migration of the nitrate 'wedge' is shown, together with a time-series of nitrate concentration for a water-supply borehole (after Foster *et al.*, 1986).

was traditionally required by the World Health Organization which set an upper limit of 100 mg l^{-1} (nitrate nitrogen) for human supplies; the European Communities limit is, however, half this at 50 mg l^{-1}. As Croll and Hayes (1990) point out, the introduction of the stricter limit increased UK water-supply costs by amounts from £15 million (e.g. for sinking new boreholes) to £199 million (to include improved water treatment).

As befits a tradition of controlling emissions, the EC has further strengthened the Community's approach to the problem by:

- Developing the same strict limits for 'raw' waters intended for water supply (i.e. pre-purification).
- Extending this control to 'sensitive' waters where cultural eutrophication is a danger to conservation, recreation or fisheries.
- Introducing the principle of protection zones where land management and use are controlled.

Under the last of these facilities the UK has around thirty *Nitrate Sensitive Areas* (NSAs); within them farmers receive financial compensation for switching from arable to grassland or trees, for reducing livestock numbers, and for planning a careful annual cycle of nutrient uptake and beneficial use rather than running the risk of leaching to surface waters and aquifers. The hydrological modelling necessary to select NSAs and to be confident that the financial outlay to farmers and reduced agricultural production will benefit pollution control ranges from simple (e.g. Cook, 1991) to very complex (e.g. Bergstrom and Jarvis, 1991)—we return to this divergence in Chapter 9.

It is necessary to make two further points to calm the 'chemophobia' engendered by the nitrate issue. One is that the leached nitrate is by no means only that added from a bag as fertilizer by the farmer. The soil is a huge reserve of nutrients and, as Tivy (1987) makes clear, traditional farming systems tend to maintain this store, at a relatively low level; a greater intensity of exploitation means less nutritional balancing, e.g. between crops and livestock, an enhanced store and more open routes to other stores such as freshwater biomass. Table 8.13(a) and (b) attempts to put the contribution from agriculture into perspective, both (i) against sources of nutrient pollution from urban/industrial sewage, and (ii) against 'natural' inputs and productive (crop) outputs.

Secondly, there is a broad danger in intensive

Table 8.13(a). Per capita output of N and P in the United States, 1965–1970*

Source	Output (kg/person-year)	
	N	P
Sewage:		
Physiological waste	4.5	0.6
Detergents	0.0	1.1
Industry	0.5	0.1
Total	5.0	1.8
Delivered to water	4.5	1.6
Agriculture:		
Animal wastes	45	6
Fertilizers	20	8
Total	65	14
Delivered to water	8	0.3

Source: Vallentyne (1974).

Table 8.13(b). Estimated inputs and outputs of nitrogen in UK agricultural land, 1978

Inputs	Nitrogen ('000 tonnes)	Outputs	Nitrogen ('000 tonnes)
Rain	275	Crops and grass	1367
Seeds	14	Leaching	326
Chemical fertilizers	1150	Ammonia volatilization:	
Sewage	26	Livestock excreta	536
Livestock excreta	1020	Crop wastes	50
Silage effluent	9	Sewage	9
Straw	15	Balance by difference	
Feed waste	9	(mainly denitrification	
Biological nitrogen		and immobilization)	380
fixation	150		
TOTALS	2668		2668

Source: Royal Society (1983).

land-use systems of cultural eutrophication—the enhanced progress of water bodies towards a high nutrient status (largely controlled in humid-temperate zones by phosphate, not nitrate). The effects of eutrophication are:

- species diversity decreases and the dominant biota change
- plant and animal biomass increases
- turbidity increases
- rate of sedimentation increases, shortening the life-span of the lake
- anoxic conditions may develop;

and the problems these raise are:

- treatment of potable water may be difficult and the supply may have an unacceptable taste or odour

- the water may be injurious to health
- the amenity value of the water may decrease
- increased vegetation may impede water flow and navigation
- commercially important species (such as salmonids) may disappear.

Nevertheless, despite the recent cases of widespread algal blooms in the UK lakes and reservoirs there is little evidence of a vastly accelerating problem, perhaps because of the controls held by phosphate concentrations which remain generally low.

8.5.3 Acidification of surface waters

The term 'acid rain' was first used during the Industrial Revolution and the effect was mainly local to industrial centres although the Norwegian playwright Ibsen referred to 'Britain's smoke cloud sinks corroding to the land' in 1866.

The local pollution was caused by rain falling through the acid, smoke-laden atmospheres of Victorian cities. As an issue it subsequently took a weak second place behind the concern generated over the direct human-health effects of the same polluted air (see Brimblecombe, 1989). The widespread desire by city dwellers to clean up the air led to increasing reliance for energy upon electricity generated out-of-town, and to large industrial plants also being located away from population centres, where the benefits of local dispersal of air pollution would be greatest (the construction of tall chimneys lofted the pollutants into a more effective air circulation—the so-called 'high-stack' policy).

The most dangerous of the pollutants from urban and industrial sources was at first considered to be *sulphur dioxide*, SO_2. However, the detection of widespread acidification of freshwater environments in the 1970s and 1980s has revealed that this form of diffuse-source pollution involves many more compounds (e.g. *nitrogen oxides*), more pathways (e.g. through algal metabolism in the oceans), and more sources (e.g. agriculture and the internal combustion engine). Principally, however, research on freshwater acidification has revealed the importance of hydrological pathways and, as a consequence, the role of land use and land management which both help to create the physical structures in the soil, affecting runoff routes, and the chemical composition of the soil which may mitigate acid deposition.

Thus, as Kirby *et al.* (1991) report for a forested catchment in Wales, which receives polluted dry deposition from the industrial English Midlands and South Wales but 'naturally acid' rainfall from the Atlantic, baseflows are well-buffered against high concentrations of hydrogen ion and the aluminium released by acidity in the soil profile. However, rain events which follow dry spells produce damagingly high 'spikes' of acidity (Fig. 8.11(a)) which are significantly greater in the forested catchment than they are for neighbouring moorland. These authors relate the land-use difference in chemical response to the scavenging capacity of the forest canopy for acid deposits and the rapid-runoff routes available via stemflow, root channels, and the drainage network. In the semi-natural catchments of similar areas of sensitive soils in Wales (Fig. 8.11(b) and (c)) the dynamic contributing-area model of runoff prevails, and in the prevalence of throughflow the albeit limited buffering capacity of these soils is better utilized. Use of the source area for lime treatment of acidified catchments is now a proven strategy.

Use of the diatom record from acidified lakes in Europe has produced the key evidence of the source–pathway–target links in the process; whilst it was always a speculation that industrial emissions were the cause, the dating of the onset of acidification (see Table 8.14) to the Industrial Revolution (only exceptionally in the list is it dated to more recent afforestation) has meant that industrial sources have been persuaded to clean-up emissions of sulphur and nitrogen oxides. This clean-up will now be carried out on a continental basis in respect of not only surface-water acidification but the *critical loads* producing impacts on a range of sensitive ecosystems. Palaeoenvironmental analysis has given optimism for the efficacy of pollution-control investments: in some acidified lakes the diatoms are already changing population numbers and diversity in response to reduced inputs of atmospheric and catchment acidity. Foresters and planning agencies in the UK have agreed on the need to prevent this key land use becoming an acidifying factor in sensitive areas (Department of the Environment/Forestry Commission, 1991).

8.6 Pollution control in the interests of conservation and recreation

Readers may have noted a paradox in the amount of attention devoted by Chapter 7 to freshwater systems as natural habitats and the emphasis given in this chapter to water use by human society. One might cynically conclude that pollution control is

Fig. 8.11. Surface-water acidification. (a) The record of an acid 'pulse' in an upland catchment. (b) An approach to mitigation by lime applications to the contributing areas. (c) The results of liming in early 1987 (all after Waters *et al.*, 1991).

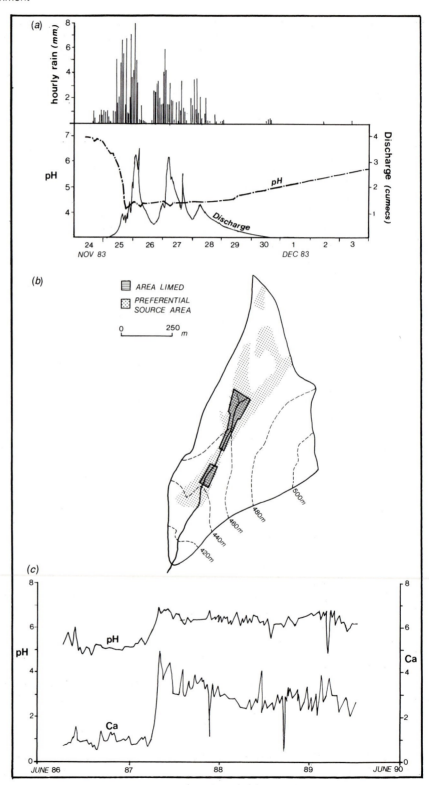

selfishly oriented to prevent direct 'own goals' in the circulation of chemicals, with interest extending to fisheries because of the property rights they confer and the fact that we eat them!

We ourselves share with other biota a 'life in water' during the many recreational activities which now occur on rivers, lakes, and reservoirs. Fewtrell (1991) expresses concern that the health risks to recreational users of inland waters are growing; recent algal blooms have resulted in the closure of recreational facilities at many reservoir sites. Sports such as canoeing may involve frequent immersion in bacteriologically polluted waters.

Furthermore, modern sensitivities about landscape and wildlife, together with scientific evidence

of the critical resource represented by biodiversity have ensured an extension of our selfish interests to new agendas of pollution control.

Two manifestations of freshwater pollution in remote rural areas in the UK have forced conservationists to consider seeking controls in the interest of pristine rivers, lakes, and wetlands: those, including acidification which result from afforestation and those, including eutrophication, which result from the intensification of all forms of land use but notably agriculture (see Newson, 1991b and c).

Catalogues of the impacts of catchment-derived pollution are now common (see Ashmore et al. 1988) and have led to calls for active intervention by government in land use—along the lines of

Table 8.14. Diatom-based pH trends for all UK sites studied

	Grid reference	Modern pH (measured)	Date of core	pH 1800 (inferred)	Modern pH (inferred)	First point of change	pH change
Wales:							
1. L. Hir	SN 789675	4.8[g]	1984	6.1	—	1850	1.3[a]
2. L. Gynon	SN 800647	5.2	1985	6.0	5.5	1945	0.5
3. L. y Bi	SH 670265	4.9	1985	6.0	4.7	1900	1.3
4. L. Dulyn	SH 662244	4.9	1985	6.0	4.7	1850	1.3
5. L. Eiddew Bach	SH 646345	4.8	1985	6.6	—	1940?	1.8[a]
6. L. Cwm Mynach	SH 678238	5.6	1985	6.1	5.6	1850	0.5
7. L. Llagi	SH 649483	4.9	1985	6.1	4.6	1850	1.5
Scotland:							
8. L. Enoch	NX 446851	4.5	1982	5.3	4.4	1840	0.9
9. L. Valley	NX 444817	4.7	1981	5.6	4.6	1860	1.0
10. Round Loch of Glenhead	NX 450804	4.7	1981	5.9	4.9	1860	1.0
11. L. Dee	NX 466788	5.3	1980	6.2	5.5	1890	0.7
12. L. Grannoch	NX 542700	4.7	1980	5.7	4.7	1930	1.0
13. L. Fleet	NX 560698	4.5	1985	5.8	4.6	1975	1.2
14. L. Skerrow	NX 605682	5.3	1980	5.9	5.8	—[b]	—[c]
15. L. Urr	NX 760845	6.8	1984	6.5	6.8	—[b]	—[c]
16. Loch Tanna	NR 921428	5.0	1986	5.0	4.6	1850	0.4
17. L. Tinker	NN 445068	5.7	1985	6.4	6.0	1850	0.4
18. L. Chon	NN 421051	5.2	1986	6.4	5.5	1900[d]	0.9
19. Loch Laidon	NN 380542	5.4	1985	5.8	5.3	1860	0.5
20. Lochnagar	NO 252859	5.0	1986	5.7	4.8	1850	0.9
21. Dubh Loch	NO 238828	5.3	1986	5.7[e]	5.2	—[f]	0.5
22. Loch nan Eun	NO 230854	5.0	1986	6.0[e]	4.8	—[f]	1.2
23. Lochan Uaine	NO 001981	5.8	1986	5.8	5.7	—[f]	—[c]
24. Lochan Dubh	NM 895710	5.6	1986	5.5	5.2	—[f]	0.3
England:							
25. Scoat Tarn	NY 159104	5.0	1984	6.0	4.6	1850	1.4
26. Low Tarn	NY 163091	5.0	1984	6.0[c]	4.6	—[f]	1.4
27. Greendale Tarn	NY 146074	5.2	1984	6.0	4.9	1900	1.1
28. Burnmoor Tarn	NY 184044	6.4	1979	6.3	6.2	—[c]	—[c]

[a] Calculated using modern measured pH
[b] Not applicable
[c] Not significant
[d] Main change after afforestation (post-1970)
[e] Refers to the base of the core
[f] Core not yet dated
[g] Pre-liming data

Source: Acid Waters Review Group (1989).

Nitrate Sensitive Areas—where diffuse-source pollutants drain into river, lake, or wetland Sites of Special Scientific Interest.

The contemporary argument surrounds *eutrophication* because of the intervention by the EC with legislation to protect sensitive waters. As pointed out in the thorough review by Henderson-Sellers and Markland (1987) eutrophication arises from processes controlling the output and circulation of nutrients in freshwater bodies, not merely the inputs; weather conditions are implicated too. The change from oligotrophic to eutrophic conditions, catalogued for lakes in Table 8.15, is one from a relatively unproductive state to a productive one—a transition which might occur naturally over a thousand years or so. The biotic components of the freshwater habitat are profoundly changed and, as a result, the physical characteristics of the water. The biggest danger to mankind is that of poor taste and colour of the water, and that from toxins emitted from the metabolism of the algae or their death and decay.

8.7 Water pollution and development in tropical and arid zones

It is simplistic but helpful to consider initially the water-quality problems of the developing world as dominantly microbiological, threatening health through disease transmission, in contrast to the chemical contamination of the developed-world freshwater system. Rapid-development of both agriculture and mineral extraction offers us, however, two examples of pollution in the developing world which spoil the simple division.

Malm *et al.* (1990) evaluate the case of gold-mining in the Amazon basin. In a major southern tributary, the Madeira River, gold-mining occurs in the river channel, and on the mining boats an amalgam is formed between the gold and mercury in order to purify the former. Between 1979 and 1985 100 tons of mercury are estimated to have entered the river system directly or via an atmospheric track (the mercury and gold are heated in the refining process). Mercury levels are high in the river sediments, fish flesh, and human hair. It is not possible to estimate the disease effects because of the endemic malaria in the region but there are contin-

Table 8.15. Comparison of main classical characteristics of oligotrophic and eutrophic lakes

	Oligotrophic lakes	Eutrophic lakes
Physical features		
Altitude	Often high	Usually low.
Proportions	Narrow and deep	Wide but shallow basin.
Substratum	Inorganic silt and stones	Organically enriched mud.
Light penetration	Good, water clear	Poor, transparency low.
Chemical water quality		
Nutrients	Poor, low in nitrogen	High, especially nitrogen and phosphorus.
Hardness	Soft	Hard.
Suspended solids	Low	High.
Dissolved oxygen	High, even on stratification	Variable, high in epilimnion but hypolimnion anoxic on stratification.
Biology		
Phytoplankton	High diversity (few individuals but of many species), low biomass; Chlorophyceae typically present, including desmids (e.g. *Staurastrum, Staurodesmus*) and Chlorococcales (e.g. *Oocystis*); diatoms (e.g. *Cyclotella, Tabellaria*) and Chrysophyceae (e.g. *Dinobryon*) occur, as well as Dinoflagellates.	Low diversity (few, abundantly represented species), high biomass; blue-green algae (Cyanophyceae) typically present (e.g. *Anabaena, Aphanizomenon*); diatoms (e.g. *Asterionella, Fragilaria, Melosira* and and *Stephanodiscus*) and Chlorophyceae (e.g. *Pediastrum* and *Scenedesmus*) also occur.
Macrophytes	Few species and rarely abundant; rare on shore but may occur at depth.	Many species, abundant in shallow margins and along shoreline.
Benthic invertebrates	High diversity but low biomass.	Low diversity, often very high biomass.
Zooplankton	High diversity but low biomass.	Low diversity but high biomass.
Fishes	Few species, characteristically Salmonidae (e.g. *Salvelinus*) and especially *Coregonus* (whitefish); low biomass.	Many species, typically Cyprinidae (e.g. *Abramis, Cyprinus, Tinca*), Percidae, Centrarchidae and Cichlidae.

Source: Hellawell (1986).

ual dangers to the river-dependent traditional lifestyle. Work in the Tyne basin, UK, illustrates the long-term effects of metal pollution of river systems. Almost a century after uncontrolled metal-mining in the headwaters there is continuing pollution via sediments deposited in the floodplain (Macklin, 1992). Remobilization of the metals into toxic forms can occur when other water-quality parameters change, particularly pH. Such a change may also be a corollary of more recent basin development.

Metal-mining often employs rivers in revealing, processing or removing waste (as was the case in the upper Tyne in Victorian times). The Star Mountains of Papua New Guinea are being exploited for gold and copper in open-cut mines in the Ok Tedi river basin. Almost 10 million tonnes of sediment is released into the river each year.

Industrialization normally follows the extractive resource phase in development and, just as happened in the developed world, it causes pollution of rivers; even if emission standards are employed, precautions against catastrophic releases of pollutants are often lax. In May 1992 a tributary of the Mekong in north-eastern Thailand was polluted by 9,000 tonnes of molasses from a sugar mill; two other aspects of development have aggravated the catastrophe: dam construction has prevented recolonisation from tributaries and the loss of fish stocks threatens the lives of subsistence communities downstream.

Almost simultaneously with the Thai episode it was revealed that 'progress' in the Nepalese carpet industry, involving semi-mechanized washing in caustic soda, sulphuric acid, and bleaching powder, has polluted both rivers and groundwater at a time of drought, promoting civil disorder in Katmandu as a result of drinking-water shortage.

One problem of cataloguing the effects of development on river-water quality is shortage of data. Those agencies attempting to improve human health in the developing world through extending pure-water supply and treating sewage, are rightly using surveillance of drinking-water quality and sanitary conditions, rather than river-water quality surveys, but now UNEP and WHO are using a network of more than 300 stations worldwide to compare water-quality parameters. The problems of access, distance, and laboratory facilities often mean that field-testing of simple parameters such as conductivity and pH is the limit of the early river water-quality surveys in developing countries; groundwater quality remains almost unexplored. Remote sensing can, as yet, provide help only with suspended sediment and certain organic products. We have already referred to the plethora of chemical determinands which may alert us to pollution; whilst reconnaissance surveys are now being made of large developing-world river systems Degens et al. (1991) illustrate how they concentrate upon sediments, carbon, and nutrients, or minerals.

There is a problem of vanishing horizons as the world sets itself increasingly stringent standards for water quality; between the WHO's 9 parameters in 1970 and its 96 parameters in 1992 there has been a huge expansion of chemical production, much of which has not yet reached the developing world.

9. Models and management

A popular response to many of the issues raised in this book, from increased urban runoff to the cultural eutrophication of lakes might be summarized by the phrase, 'Something must be done'. Hydrology has always been an *applied* science; its environmental connections and relevance become clearer year by year and therefore it should and can now begin to make a contribution to the high levels of *global environmental management* which we know are inevitable. There are in fact two component questions asked of a body of knowledge in the position now occupied by hydrology; the first is to convert the urgent popular view into a question: 'Must something be done?' It is the responsibility of all environmental scientists to enable society to interpret future trends as reliably as possible without recourse to hyperbole. The second question is, of course: 'How do we do what must be done?'

9.1 What are predictive models?

One of the major channels in which applied sciences operate is *prediction*. Formal knowledge of the present and/or past is used to anticipate the future development of the phenomenon under study; the natural environment is, of course, a multi-component phenomenon, best understood as an *interactive system* but, nevertheless, difficult to both understand and predict. Fig. 9.1 puts, very generally, the role of prediction and its two main forms—

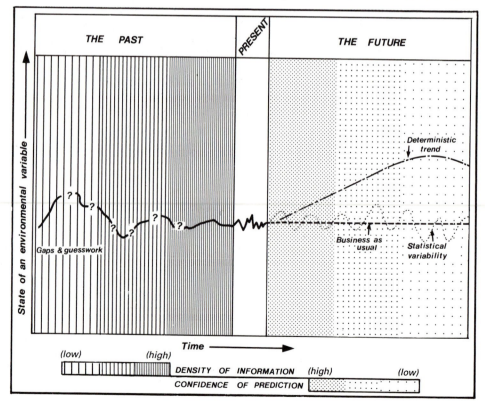

Fig. 9.1. Sources of information for, and scientific approaches to, prediction. The context of modelling for management.

deterministic understanding of the operation of a system and *statistical* description of current states and future trends. Often the statistical approach is forced upon us by a richness of data but a relative poverty of understanding; we are forced to let the numbers speak for themselves rather than 'acting God', the position of deterministic predictions. One advantage of statistical techniques is their facility to account for random processes and events which we know to be common in environmental systems—*stochastic modelling* allows these random elements a formal role.

Whatever technique of prediction is employed we can apply the general term *model* to our abstraction of complex reality; even the saying 'red sky at night is a shepherd's delight' is a form of model. Like almost all models it is flawed because, even if we know what makes shepherds delighted, we cannot invariably match these conditions to a red sky at night—our predictions are wrong at least part of the time. Models, to be of most value, must quantify their vulnerability to failure—we need *confidence limits* which we can apply to usable predictions (see Fig. 4.6). Internal analysis of the predictive technique is required to identify the *sensitivity* of different components—their relative effect on the outcome compared with the effects of the other components.

One of the advantages of modern data-gathering (e.g. relaying river-flow data by telemetry or satellite remote-sensing of rain clouds) is that predictions can be updated in real time 'as the future arrives'. We can revise our prediction of a flood peak as the causative rain storm develops or disperses. The most sophisticated of such systems in real time allows us to *forecast* not only the state of the future environment but its time frame and variability through time.

Flood prediction and flood forecasting (see Chapter 4) are at one extreme of the demands made on hydrology by society; Table 9.1 indicates that at another—represented by the flow of rivers in AD 2030, following climatic changes resulting from the enhanced 'greenhouse effect' we can do little better than set scenarios, asking simple 'what if . . . ?' questions, and building into the answers such imponderables as society's own responses to the problem. Science is now acting in a new role (in relation to society) with predictions which, though necessarily imprecise, are being used to guide management under a *Precautionary Principle*.

Table 9.1. Development of scenarios of UK river runoff consequent upon global warming

Catchment	Present evaporation changed precipitation (%)	Present precipitation changed evaporation (%)
Greta	+ 29	−7
Dove	+ 31	−7
Wensum	+ 40	−12
Lambourn	+ 58	−25
Avon	+ 42	−14
Tamar	+ 32	−8
Teme	+ 38	−12
Cynon	+ 27	−5
Eden	+ 27	−5

Note: The table selects the maximum range of estimates for each catchment from seven scenarios and rounds up the estimates to the nearest integer percentage.

Source: Arnell *et al.* (1990).

9.1.1 A 'model of models' in hydrology

Few readers will devise or operate hydrological models or their derivatives used in pollution control, conservation, or other fields of environmental management. Most will, however, need to understand a brief typology of hydrological-modelling approaches and the potential and limitations of each type; the reason is that modelling increasingly underlies our management targets as computing power and data-gathering capacity expands.

The field of hydrological modelling may be defined in terms of the wide range of inputs, modes of operation and outputs. The model is chosen or constructed (few are available 'off the shelf' and modellers rather jealously defend their work) after a compromise between aims, limits of theoretical knowledge, available data, and operational practicability. Two main problems immediately suggest themselves: an academic research-based model with a generous input data supply may end up with no users, and a model originating from a highly specific application may be threatened by data shortage. Fortunately the rapid rate at which hydrological databases are now expanding through *remote sensing* means that the era of data-limited modelling is coming to an end. Models are nevertheless greedy for data now that computing power has expanded to the point where manipulation of data and algorithms is possible on quite small machines (Box 9.1).

9.1.2 Two models compared

One of the basic divisions between types of deterministic hydrological models is that based upon the ability of the predictions to incorporate *spatial variability* within the catchment for which the model is being *calibrated* (i.e. fixed to be a good representation of reality). The two resulting classes of model are labelled *lumped* if they represent the catchment as a single unit or *distributed* if they incorporate the spatial variety of rainfall, snowpack, soils, slopes, or channels which we know to be present in nature.

The lumped catchment model shown in Fig. 9.2(a) is that described by Blackie and Eeles (1985); the authors preface their work with a plain statement of their approach:

It is implicitly assumed that the objective of models is to obtain a representation of the streamflow emerging from the catchment.

A model, like a painting or a sculpture, is a representation of reality.

When the natural system is as complex, large and imperfectly understood as a catchment it is unlikely that a complete representation of every process occurring at every point in the system can ever be achieved.

Thus the aim must be to identify the major processes contributing to the response that is of particular interest

. . . until the best possible results are achieved within the constraints of time, computing power and the modeller's own ability and experience. (pp. 311–12)

Blackie and Eeles see their model (Fig. 9.2(a)) as extending a tradition of simple input–algorithm/equation–output devices available in hydrology since the 1850s. In their case, however, the major internal functions of the catchment are separated after the fashion of storage tanks (in the manner we have developed throughout this book). The model works with hourly inputs of precipitation, modified by hourly values for potential evapotranspiration from climate data. The model's tanks are described by fifteen *parameters* or numerical descriptions of the capacity of the tank, its rate of overflow, and time delays in flow between tanks. At each stage of the model the parameters (abbreviated in Fig. 9.2(a)) are linked by simple equations as the volume and timing of the rainfall cascades through the tanks.

Once set up and working well on the computer this kind of model must be calibrated against some real catchment data, i.e. where stream discharge is known in addition to rainfall and evapotranspiration. The internal workings and parameter values of the model can then be set so as to get the best and most consistent result—they are *optimized* by a well-established set of trial-and-error mathematical techniques. Perhaps half the 'real' data are used to set up the model and it is then run in predictive mode to see how well its streamflow simulations compare with the rest of the record.

Clearly such models require some data to exist already and are therefore at their most useful in checking the accuracy of gauged data, extending short records to investigate climatic change and estimating the bulk effects of a land-use change.

For student use, the computer model of this type supplied by John Wiley & Sons under the title of 'RUNMOD' is a very useful tool (Prowse *et al.*, 1986). A much more varied set of models is described for student use (and made available on disk) by Kirkby *et al.* (1987; 2nd edn., 1993).

In contrast the set of distributed models described by Beven (1985) and the example shown in Fig. 9.2(b) have, as their aim, the best possible mathematical description of the processes operating in all parts of a catchment; these models do not require calibration against a real record but do require faith in hydrological theory together with, in some cases, a large field-work component to check the workings of the modelled processes. Beven makes his philosophy as clear as that of Blackie and Eeles:

Fig. 9.2. Hydrological models. (a) The lumped parametric type (after Blackie and Eeles, 1985). Lettering refers to parameter names. (b) The physically-based distributed type (after Beven, 1985).

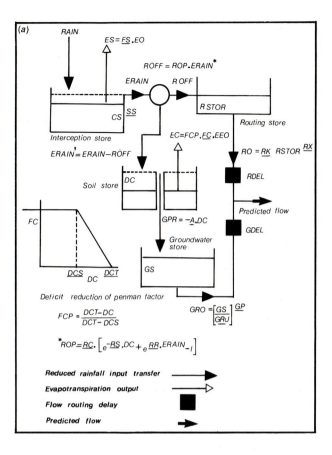

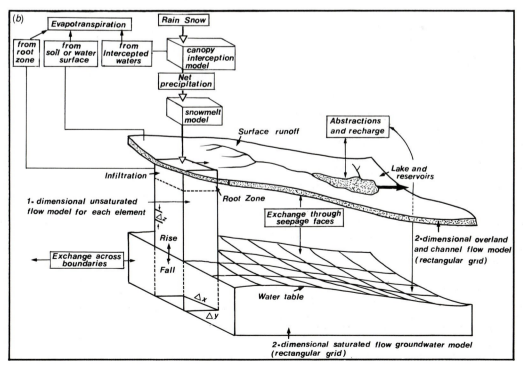

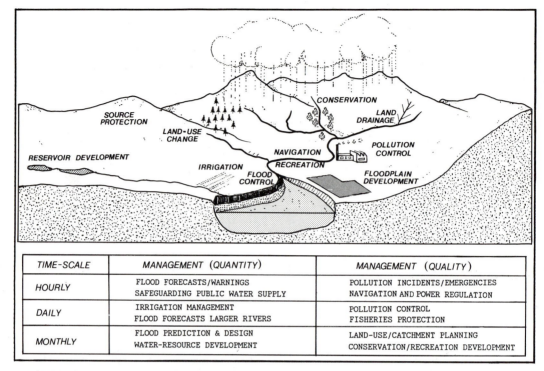

Fig. 9.3. The aims of river basin management at various time-scales and as distributed throughout the catchment system.

The following table appears within the figure:

TIME-SCALE	MANAGEMENT (QUANTITY)	MANAGEMENT (QUALITY)
HOURLY	FLOOD FORECASTS/WARNINGS SAFEGUARDING PUBLIC WATER SUPPLY	POLLUTION INCIDENTS/EMERGENCIES NAVIGATION AND POWER REGULATION
DAILY	IRRIGATION MANAGEMENT FLOOD FORECASTS LARGER RIVERS	POLLUTION CONTROL FISHERIES PROTECTION
MONTHLY	FLOOD PREDICTION & DESIGN WATER-RESOURCE DEVELOPMENT	LAND-USE/CATCHMENT PLANNING CONSERVATION/RECREATION DEVELOPMENT

distributed models will be taken to mean models of catchment hydrology that are physically based . . . firmly in our understanding of the physics of the hydrological processes which control catchment response. Physically based models are necessarily distributed because the equations on which they are defined generally involve one or more space co-ordinates. They thus have the capability of forecasting the spatial pattern of hydrological conditions within a catchment as well as simple outflows and bulk storage volumes. (p. 405)

The equations about which Beven writes are those which we have described earlier in this text relating, for example, to flow in the saturated zone (p. 40) and in open channels (p. 21). They are generally phrased for modelling as non-linear partial differential equations (cf. the simple numerical operations of the Blackie and Eeles model); solutions must be found for these equations at nodes in space and time. As Fig. 9.2(b) shows, the resultant discretization of the catchment is most frequently one into grid squares, an arrangement particularly practicable for computers and for data derived from remote sensing.

Because distributed models are expensive in both computing and field-work their justification must be in terms of end uses; four are particularly relevant:

- predicting the results of land-use change
- predicting spatial variability of response from, for example, localized rainfall
- predicting the movement of sediments/pollutants
- predicting/forecasting the flow from managed catchments.

One of the most satisfying aspects of distributed models is their ability to accommodate the spatial variability of runoff processes (Troendle, 1985) which we made a feature of in Chapter 3 and to whose description geographers have made such a major contribution (Newson, 1985). Nevertheless, some form of incorporation is also possible in lumped models by careful use of 'tanks'.

9.1.3 How good are the models?

The acid test of a model, like that of an unveiled portrait or sculpture (Blackie and Eeles' analogy) is that of how well it compares with reality. If it is not an exact replica are its shortcomings or embellishments at least a contribution? Models may be said to have *heuristic value*—we learn by constructing or using them as an academic exercise.

The formal test of a hydrological model is to plot observed and predicted values of streamflow, sediment concentration, pollutant load, etc. A goodness of fit index is also calculated. Taking the outflow stream discharge as an example, the test may be event-based, i.e. one storm hydrograph, or may be continuous over a much longer period. Tests of the latter type often allow the modeller to update or refine the predictions with the developing actual conditions, perhaps relayed by telemetry or satellite from the catchment to the river manager.

Our need to predict and forecast over long periods in the future to protect against unforeseen impacts of pollution of climate change has produced a new sort of test for models in recent years—that of replicating known trends from the past. The simple lumped model of stream acidification, MAGIC, has derived much of its following in hydrology from its success in 'retradicting' (cf. predicting) the acidification record as witnessed by palaeoecological studies of lake sediments (see Section 8.3.2).

The current progress in modelling hydrological and related systems is summarized in Table 9.2. It should be emphasized, as a link with Section 9.2, that many if not most models are derived for, or calibrated on, small catchments which may be relatively remote from where predictions are needed for river-basin management. For some purposes the answer to this problem is to use the outputs from models for small areas as inputs to routing models which account for travel within the river network downstream and for the accompanying processes of storage, dilution, and dispersion.

9.2 Principles of river management

Many authors have recently set down the challenges faced by those responsible for managing the resources of river basins, not merely the immediate ones of resource use but the longer-term challenges encompassed by the word sustainability. These problems are especially challenging in the world's largest river basins which cross international boundaries, e.g. the Nile, with nine contributing nation states. The river basin is an environmental unit, brought together by gravitational processes, but within it there is normally a good human perception of belonging. The major intellectual challenge appears, therefore, to reconcile the hard skills of the scientist and engineer with the softer aims of sociologists, politicians, and planners (the whole then wrapped in a legal and economic framework).

The aims of sustainable river basin management are set out in Fig. 9.3. However uplifting the fine words of this agenda may seem they must be translated into an operational set of management activities which refer to both shorter and longer time periods (Fig. 9.3). The separation of quantity and quality aims is artificial but encourages a link between quality of water and quality of life, through activities such as recreation and conservation.

9.2.1 River-basin management and environmental protection

It requires a considerable sophistication in public policy to move from consideration of water as a resource to be exploited, and a hazard to be controlled, to a position where the considerable breadth of environmental coverage of this book becomes the integrated agenda for management. Whilst geographers, biologists, and other environmental scientists might be able to see the virtue of holistic treatments, engineers and politicians might be more conservative. It is clear, however, that the previous hydraulic age is metamorphosing into a hydrologic age since human needs and numbers require a detailed understanding of the collection and transfer processes in river basins as well as the traditional distribution systems of canals, pipes, and drains (Newson, 1992).

What do we mean by environmental protection? It becomes an operational concept if we divide

Table 9.2. Current trends in hydrological modelling and constraints/opportunities

Model purpose	Theory?	Data?	Widespread adoption?
Channel flow/ flood routing	Good	Yes	Yes
Flow from catchment (extremes)	Fair	Growing	For complex basins
Flow from catchment (normal)	Good	Fair	Need to incorporate land use
Sediments	Fair	Poor	Virtually no uptake
Solutes/ pollutants	Fair	Poor	Rapid growth/uptake: simple, point-source models

between *preservation* and *conservation*. Preservation in river basins may mean the protection of ancient archaeological remains or even the water industry's own history in the form of pumping stations. It may require relocation of artefacts during, for example, reservoir construction. Seen more broadly it can incorporate the mitigation of short-term hazards which might, in a natural system, mark the beginning of a threshold change to conditions which humans would consider as degradation. Conservation has many interpretations but at its heart refers to the maintenance of sufficient past information for future uses, be that information genetic or in the form of the environmental record buried in the peat of a bog or the sediments of a lake.

It is therefore essential that river managers do not equate environmental protection purely with pollution control. However, one need look no further for the long-term benefits of pollution control than its role in achieving sufficient freshwater supplies for the globe's burgeoning population (see Chapter 1). Fig. 9.4(a) indicates the sharp rise in this century in the volume of river water which is polluted by waste water; the graph is optimistic in forecasting a reversal in the trend. The fight to pre-

vent groundwater from a similar fate in future must take a high position in the longer-term view of pollution control. Fig. 9.4(b) illustrates the ways in which our measurement, research and education in hydrology need to interface with legal or economic processes of bringing about change; our hydrological 'discoveries' do not become the basis of public policy by simple diffusion!

9.2.2 Conservation, enhancement, and restoration of the wet environment

It must be admitted that our treatment of river-basin systems to date has shown too many examples of lack of understanding of those patterns and processes which control the natural operation of their resource base in the longer-term and at the larger scale. The heroic engineering talent for 'problem solving here and now' has been partly to blame but we have all been grateful for those problems to be solved. It is conventional for environmentalists to concentrate their attacks on the shortcomings of river-basin development in the developing world (e.g. Goldsmith and Hildyard, 1984, 1986, 1992) but the mistakes are equally apparent in the developed world.

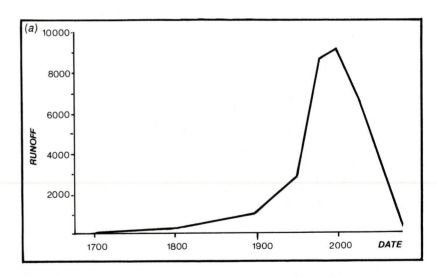

Fig. 9.4. The essential control of freshwater pollution. (a) The component of waste water in global surface flows (separated from Fig. 5.12) (after L'vovitch and Gilbert, 1991). (b) Scientific and policy processes in pollution control.

As a consequence we may introduce two new words beside the conservation theme for sustainable development. They are *enhancement* and *restoration*. Quite simply we should judge further river-basin developments on the basis of how they contribute to conservation as well as to exploitation. Very often the extra financial costs of enhancement are tiny compared with those of the main scheme. For example, can a flood-protection scheme benefit a major wetland by shedding to it the surplus flood water? Can sewage treatment occur by carefully zoning the agricultural land use of floodplains and by developing artificial reed-beds?

Because land drainage and flood protection in the humid temperate lands is an extensive and manipulative form of development and because of the sheer extent of channelization (see Chapter 5) it has been this sector of engineering hydrology which has made the most promising reforms in recent years. The environmentally sympathetic alternatives to conventional hard engineering are shown in Table

9.3; they include stream restoration and renovation—Fig. 9.5 shows how the former process can undo the worst excesses of 'river training' formerly thought to be the best form of flood and erosion control in the developed world. We are heading back towards the prehistoric Chinese engineering philosophy of 'put nothing in the river'.

9.3 The context of management: science, people, and planning

No matter how successfully we incorporate our knowledge of patterns and processes into predictions in hydrology and the river environment, we need an interface with society whereby effective use can be made of both technical and non-technical insights. Developed nations are still experimenting with formal interface systems with the following aims:

Table 9.3. Examples of alternative engineering practices

Guidelines	Area to which applied	Source
Pools and riffles Construction of pools and riffles provides stable-channel morphology that is biologically productive and aesthetically pleasing.	General application in North America	Keller (1975)
Rest stops Concreted channels alternated with short lengths of natural channel which provide an acceptable habitat for fish. Permits fish migration in an otherwise impossible situation.	Hawaii	Parrish *et al.* (1978)
Stream restoration Minimal straightening of channels; retention of trees to promote bank stability; minimization of channel reshaping; use of bank stabilization techniques; emulation of the morphology of natural stream channels.	Briar Creek, Charlotte, North Carolina, employed since 1975	Keller and Hoffman (1976) Nunnally (1978)
Stream renovation Similar to restoration, but water-based methods of channel maintenance used or small tracked vehicles. Hand labour preferred.	Wolf River, Tennessee	McConnell *et al.* (1980)
Bio-technical engineering Emulation of the morphology of natural channels with meanders and asymmetrical profiles; preservation or creation of natural habitats for flora and fauna; bank stabilization with living vegetation	Bavaria, Germany	Binder (1979) Binder and Grobmaier (1978) Binder *et al.* (1983)
Meander restoration Restoration of a straightened channel to original course. Re-creation of width/depth, substrate, slope.	Jutland, Denmark	Brookes (1984, 1987*a*)

Source: Brookes *et al.* (1983).

Fig. 9.5. The reversible roles of engineering channel 'improvements' (a) which tend to damage river habitat, and restoration (b) including reconstructed natural channel features.

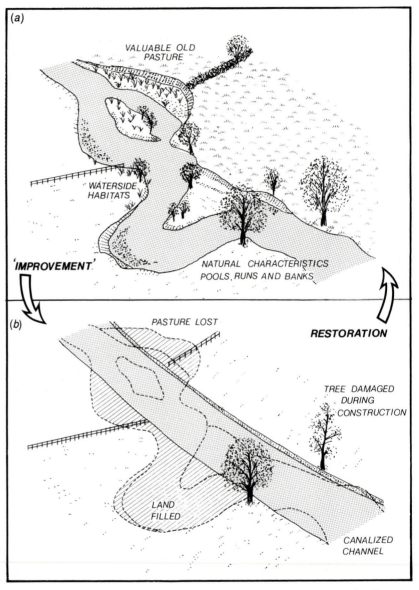

- To anticipate the impacts of a development or management device before it is accepted/implemented.
- To test alternatives in terms of their benefits and costs (with the latter including a broad *scoping* of environmental values as costs).

These are applied to proposed developments in the form of Environmental Impact Assessment (EIA) and Cost-Benefit Analysis (CBA) respectively. However, a large number of environmental issues occur outside the field of large development projects. We need, therefore, to support those concerned with

more general *policy development* and with *planning*. Those operating in such fields require the use of one or more of the models we have described but in a context of aiding a wise decision. Thus, alongside the use of predictions in EIA and CBA, models are used in computer-based Decision Support Systems (DSS). DSS can be made available to public consultation in the spirit of democratic decision-making.

Whatever our power to anticipate impacts, determine the best alternative, and aid decisions, sustainable environmental management requires the monitoring or auditing of outcomes. Models are not the antithesis of field-work—the approaches are complementary. Monitoring is by no means out-moded: its time has come and it is getting cheaper. It is decisions based upon both modelling and monitoring which we need to improve. The term 'hydropolitics' has been derived for the Nile basin's complex web of water-resource problems between the nine nations whose boundaries overlap with those of the river. Hydropolitics will, however, envelop us all in the next half-century as the continuing momentum of human development occurs against a background of accelerating environmental change. Table 9.4 attempts an assessment of the development process in terms of intervention in the water cycle. The challenge to hydrology and to the protection of the wet environment must be to anticipate and to aid good decision-making. The principles of pre-historic hydraulic civilizations have little to offer when the water to be distributed is 'too much, too little or too dirty'—the principles of a hydrologic age must be defined and empowered by research and education.

9.4 Hydrology, environmentalism, and global futures

Final work on this book is being done to the accompaniment of news from 'The Earth Summit', the United Nations Conference on Environment and Development, in Rio de Janeiro. Water supply for human consumption, industrial and agricultural use, and for the generation of power, is a prominent issue at the Summit. Glacken (1956) reveals the part played in classical times and in the earliest days of modern environmentalism by water issues, mainly linked to forest exploitation. Plato linked loss of forests to flooding, soil erosion and degradation of the countryside.

Now once again hydrological issues are at the heart of concepts of sustainable development and the conservation of biodiversity, both of which have to be set against a changing set of climatic controls. Given the central concern in world development for human health and well-being the imbalance between human numbers and needs is at the core of concerns in hydrology; water is a major limiting parameter to sustainable and healthy development, not only as a resource, but because of the hazardous extremes which it displays in the 'natural' cycle. Falkenmark (1989) has suggested that, in the field of food production, we in future should measure *production per unit volume of water* rather than per unit area of land, as a measure of the intensity and efficiency of resource use in development (Fig. 9.6).

Shiklomanov (1989) makes predictions about the amount of water required for the human cycle to the year 2000; Table 9.5 also illustrates the rapid growth in usage this century. Shiklomanov divides his data between total use (much of which is recyclable, as with human supplies) and 'irretrievable' use (leading to evaporative 'loss'). The table indicates that water supplies will be exhausted in terms of the direct line to resources but of course techniques such as storage and purification will clearly

Table 9.4. Stages of human intervention on the waters of the world

Continent	Eighteenth century	Nineteenth century	Twentieth century
Africa	No impacts on regional scale.	First stage: probably only minor local impacts.	Second stage in some major basins.
Asia	First and second stages: concentration of local population and irrigation.	Second stage: large impacts from locally dense population and irrigation.	Second stage continuing.
Australia	No effects.	First stage: probably only local impacts.	Third and probably fourth stage with planning for compatible development.
Europe	Second stage: health effects are the major.	Mostly second stage: due to increasing population and the Industrial Revolution. Third stage improvements near the end of the century.	Third stage throughout the periods around the world wars; fourth stage strongly indicated.
North America	First stage: limited, local.	Second stage: population and industrial growth: little recovery in the last years.	Third stage throughout: fourth stage initiated in many areas.
South America	No significant impacts on a regional scale.	Second stage in some areas, no impacts in others.	Second stage in many large areas; some third stage near urban centres.

Source: L'vovitch and White (1991).

Fig. 9.6. The hydrological 'limits to growth' (after Falkenmark, 1989). (a) Vegetation and crop (food) limits, (b) Population limits.

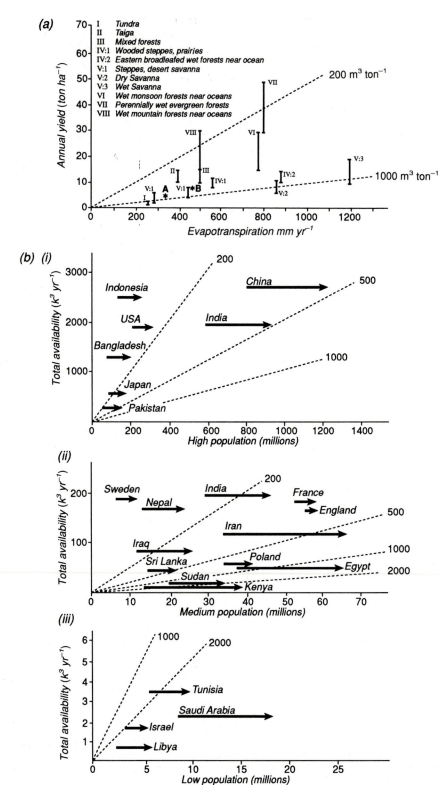

Table 9.5. Human water usage in the twentieth century

Continent	Mean annual runoff (mm)	Consumption (km^3) (total/irretrievable)			
		1900	1950	1990	2000
Europe	310	38/18	94/38	555/178	673/222
N. America	340	79/29	286/107	724/255	796/302
S. America	640	15/11	59/45	150/87	216/116
Africa	150	41/34	56/44	232/165	317/211
Asia	330	414/323	860/654	2440/1660	3140/2020
Australia/ Oceania	270	2/1	10/5	38/17	47/22

be vital; increased evaporative loss from irrigation may total 3000 km^3 per annum by the year 2000. This extra moisture will become part of the global circulation but very little of it will be reclaimed as precipitation within the region of 'loss'.

We will return later to the influence of likely climate change on this pattern of consumption and availability but first we turn to the issue of the environment.

Marchand and Toornstra (1986) demarcate the essentials of an ecological basis to river-basin development, including the need to differentiate between the intensive and extensive resources of each basin. It is the extensive resources, formed by the interaction of the hydrological cycle and the physiographic system of the drainage basin which are now being urgently identified, calibrated, and protected.

To hydrologists the increasing interest in conservation alters the focus of attention from measurements of rainfall and river flow, foremost in the era of engineering exploitation, to storage and evaporation which control the occurrence of wetlands and account for many of the effects of land-use change on the hydrological resource. As will be obvious from Chapters 3 and 5, only the major processes have been identified and measured for each major land-use modification (e.g. surface-flow alterations following urbanization). We are left to discover the more subtle and perhaps longer-lasting aspects of artificial influence on the hydrological cycle. Hydrological-measurement networks may need to be rebalanced in future; for example, networks of soil-moisture measurement will inevitably be a feature of the expansion of irrigated agriculture. Inevitably, remote sensing will expand its influence on measurement systems.

As an example of the positive utilization of known and predictable land-use effects the Australians are becoming successful in planting eucalypts to reduce the seepage of saline groundwater into agricultural areas of Western Australia (George, 1990; Schofield, 1990). More than half of the state's recoverable surface waters are saline due to this problem, partly as a result of deforestation to create arable land. By planting trees on as little as 2 per cent of a problem catchment they are able to reduce groundwater tables below existing seepage lines; eucalypts grow rapidly—a 6.5 m high plantation established in three years reduced the water-table by more than a metre and facilitated the growth of wheat on land which had been unproductive for 20 years.

Ecologically acceptable interventions of this type in the basin hydrological cycle are also increasingly likely in the name of stream restoration. Featureless but efficient drainage channels, designed to speed the runoff to the sea, can be restored by the reintroduction of meanders, riffles and pools. They can be integrated with riparian wetlands formed by 'setaside' land from agriculture. Petersen et al. (1992) describe a building-block model for this form of restoration and quote nitrogen reductions of 68–100 per cent in runoff passing through such buffer strips before entering suface channels. In this way water quality is improved by restoration of the filter/stripping function of the natural system. Other forms of physical rehabilitation include the creation of instream habitats with the addition of weirs, riffle and pool sequences, and the re-creation of sinuosity (Gore, 1985; Brookes, 1992)

The river system is inherently robust. Even the Rhine, once a salmon river but now a polluted ship canal, shows suprising resilience. Lelek and Kohler (1990) describe the recovery of the system after the Sandoz (Switzerland) pesticide pollution episode of 1986; whilst apparently 'wiping out' the life of the river all existing taxa have recovered. There is, however, a long-term reduction of diversity by seven species and it is the ambition of the Rhine Action Programme, prompted by the Sandoz incident, to make the river a salmon fishery again. Broer (1991) describes a system of water-quality alarm stations which will help patrol the river; 50 million people live in the 185,000 km² basin— imagine a salmon river draining such a catchment area! Inspiration may be gained from the success of restoring salmon to the Thames and the Tyne.

9.5 Hydrology and climate change

It is clear that hydrologists must play a leading part in the scientific response to the problems of predicting the direction, magnitude and effects of climate change (Fig. 9.7). Askew (1991: 392) reports the words of the Swedish Environment Minister, Mrs Birgitta Dahl:

As to water conditions, impacts are to be feared on precipitation, evaporation, water storage in soil, snow and glaciers, water temperature, water quality and the sea-level. These impacts will change the conditions for hydropower production, agriculture, forestry, fishery, drinking-water supply, urban hydrological facilities and harbours and will of course influence the economy.

Arnell *et al.* (1990) have particularized the sensitivities of water management issues in the UK to various components of climate change (Table 9.6).

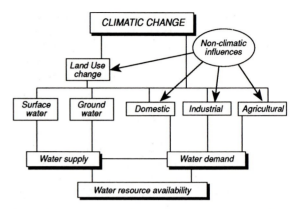

Fig. 9.7. Influences of climate on water resources (after Arnell *et al.*, 1990).

Askew lists the position of both pure and applied hydrology as follows, between the role of the climatologist and the sociologist/economist:

Climate change
Hydrological processes
Water Resource systems
Society

For this reason hydrology plays a central part in the current World Climate Programme and has a lead role in the GEWEX scheme (Global Environment Water Cycle Experiment) which attempts to connect up ground-level hydrological predictions to the

Table 9.6. An initial assessment of the sensitivities of water management issues in the UK to climatic change

Water management activity	Dimension of climate change				
	'Hydrological'			Increased temperature	Higher sea-level
	Total runoff	Seasonal distribution	Extreme behaviour		
Surface water supplies ('small storage')	•	••	••	••	o
Surface water supplies ('large storage')	••	•	••	•	o
Groundwater supplies	••	••	••	•	•
Agricultural water use	••	••	••	••	o
Flood management	o	•	••	o	••
Urban stormwater drainage	o	o	••	o	•
Reservoir safety	o		••	o	o
Drainage of low-lying lands	••	••	••	o	••
Power generation	••	••	•	o	o
Water quality	•	••	••	••	o
Effluent dilution	•	••	••	•	o
Inland navigation	•	••	o	o	o
Fisheries	o	••	•	••	o
Recreation	o	•	o	•	o
Saline intrusion into estuaries	•	•	o	o	••
Aquatic ecosystem management	o	••	••	••	o

Key: o = little sensitivity; • = some sensitivity; •• = very sensitive.

Source: Arnell *et al.* (1990).

global-climate models which are the main basis for the scenarios of climate change so far published. For example, the experiment hopes to link runoff patterns in the Mississippi basin to remotely sensed atmospheric patterns above; the same is proposed for Europe, in this case as a grid-square configuration.

At present hydrologists are torn between the use of analogues and the output of General Circulation Models (GCM's) in predicting the impacts of global warming over the next half-century. The GCM's are, by name, general—their spatial resolution may be as coarse as 500 km because they seek to model the planetary circulation. The use of analogues refers to the identification of warm and cool periods within the climate record of a region or basin and assumes that future trends will follow those of past warm spells. In Britain, however, these sources of evidence produce conflicting results (Arnell et al., 1990). Combining the evidence from both sources leads to the very preliminary conclusion that Britain may expect drier summers and wetter autumns with the increase in precipitation being most notable in the north and west. Such predictions, however, make no comment on the intensity or duration of precipitation, which will depend on the reaction of 'normal' synoptic weather patterns to climate change. Newson and Lewin (1991) indicate the difficulties of predicting the response of floods and droughts to climate trends; they conclude that floods are characteristic of the north during periods of warming. Britain is a useful geographical unit to illustrate the problems of the impact of climatically forced hydrological changes because of its small,

extremely variable topographic surface, and the dependency of its water-resource and land-use systems upon steep gradients of rainfall, runoff, and population density.

Whilst early attention to hydrological predictions for an era of climate change has focused on precipitation, the most intriguing prospects concern evapotranspiration. Changes in this variable are likely to involve the concentration of carbon dioxide, the temperature, and the land use. Stomatal conductances tend to reduce as CO_2 concentrations rise, but the effects vary with crops and there may be a compensatory rise in leaf area, e.g. as the result of warmer, wetter conditions in the British case, remembering that we are here discussing potential evapotranspiration and that choice of crops and other aspects of land use may determine whether there is a plant response.

Perhaps the most serious implications of climate change will be in those areas we call drylands. The process of development brings to these hazardous parts of the Earth an initial need for irrigated agriculture to support population growth and exportable surpluses but thereafter urbanization and industrialization. Both processes raise the demand for water but also reduce the supply through pollution, sedimentation of reservoirs, or exploitation of non-renewable groundwater. In many semi-arid lands the 'mining' of groundwater recharged during previous eras of moist climate gives cause for grave concern; for example, the use of groundwater in Saudi Arabia will shortly reach more than twenty times the annual recharge under present climate. The urban populations of these lands may well need

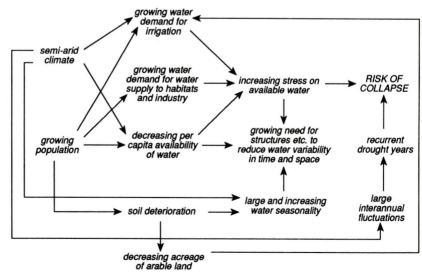

Fig. 9.8. The dangers of water shortages in semi-arid developing lands (after Falkenmark, 1989).

to forgo the swimming pools and fountains of the American 'sun-belt', together with the privilege of pollution. The rural irrigators must use water more efficiently and this response is likely to be forced on them by competition from the urban market.

The ability to forecast drought in these areas is going to be crucial (Fig. 9.8). However, such forecasts are of little use without a political translation into action, or at least awareness by governments. Malin Falkenmark's (1986,1989) proposal extends to a consideration of a hydrological carrying capacity (people per flow unit). She estimates that 2000 people per million m^3 yr^{-1} is the technical limit; at 500 there is a barrier when water-based development becomes a constraint. There is a water barrier—engineers have not given this impression.

Remote sensing is also likely to have an increasing role in the recovery of surface hydrometric data, such as rainfall and streamflow. Hard-wearing sensors which will operate untended for long periods and which send data via satellite telemetry will play an increasing part in extending our knowledge of the components of the land phase of the hydrological cycle.

Askew (1991) concludes:

Global scale hydrology cannot be ignored and it must be pursued with rigour and supported with adequate resources. If the hydrological community is too apologetic and over cautious it will be doing a disservice to the cause of international co-operation and it will certainly not fulfil its obligations to the world at large. (p. 399)

References

Chapter 1

Bertalanffy, L. von (1956), 'General systems theory' *General Systems Yearbook*, 1: 1–10.

Falkenmark, M. (1989), ' The massive water scarcity now threatening Africa: Why isn't it being addressed?', *Ambio*, 18/2, 112–18.

Keller, R. (1984), 'The world's fresh water: Yesterday, today, tomorrow,' *Applied Geography and Development*, 24: 7–23.

Lough, J. M., Wigley, T. M. L., and Palutikov, J. P. (1983), 'Climate and climate impact scenarios for Europe in a warmer world', *Journal of Climatology and Applied Meteorology*, 22: 1673–84.

L'vovitch, M. I. (1973), 'The global water balance', *United States International Hydrological Decade Bulletin*, 23: 28–42.

—— (1979), *World Water Resources and their Future* (American Geophysical Union: Chelsea, Mich.).

—— and White, G. (1990), 'Use and transformation of terrestrial water systems', in B. L. Turner *et al.* (eds.), *The Earth as Transformed by Human Action* (Cambridge) 235–52.

McBean, G. A. (1989), 'Global energy and water fluxes', *Weather*, 44/7: 285–91.

Nace, R. (1974), 'General evolution of the concept of the hydrological cycle', in *Three Centuries of Scientific Hydrology* (anon. ed.) (UNESCO: Paris), 40–8

Thom, A. S., and Ledger, D. C. (1976), 'Rainfall, runoff and climatic change', *Proceedings, Institution of Civil Engineers*, 2/61: 633–52.

Ward, R. C. (1975), *Principles of Hydrology* (McGraw-Hill: Maidenhead, UK).

Chapter 2

Bathurst, J. C. (1988), 'Velocity profile in high-gradient, boulder-bed channels', *Proceedings, International Conference on Fluvial Hydraulics* (Budapest, Hungary), 1–6.

British Hydrological Society (1988), *Weather Radar and the Water Industry*, Occasional Paper 2 (Wallingford, UK).

Browning, K. A., and Hill, F. F. (1981), 'Orographic rain', *Weather*, 36/2: 326–9.

Calder, I. R. (1990), *Evaporation in the Uplands* (John Wiley & Sons: Chichester, UK).

—— and Rosier, P. T. W. (1976), 'The design of large plastic sheet net rainfall gauges', *Journal of Hydrology*, 30: 403–5.

Ford, E. D., and Deans, J. D. (1978), 'The effects of canopy structure, stemflow, throughfall and interception loss in a young Sitka Spruce plantation', *Journal of Applied Ecology*, 15: 905–17.

Francis, J. R. D., and Minton, P. (1984), *Civil Engineering Hydraulics*, 5th ed. (Edward Arnold: London).

Haggett, C. M. (1989), 'Weather radar for flood warning', in *Weather Radar and the Water Industry* (British Hydrological Society: Wallingford, UK).

Hey, R. D. (1979), 'Flow resistance in gravel-bed rivers', *Proceedings of the American Society of Civil Engineers*, 110: 1519–39.

Hudson, J. A. (1988), 'The contribution of soil moisture storage to the water balances of upland forested and grassland catchments', *Hydrological Sciences Journal*, 33/3: 289–309.

Institute of Hydrology (1981), *User's Handbook for the Institute of Hydrology Neutron Probe System* (Institute of Hydrology: Wallingford, UK).

Jackson, I. J. (1989), *Climate, Water and Agriculture in the Tropics*, 2nd ed. (Longman: Harlow, UK).

Kirby, C., Newson, M. D., and Gilman, K. (1991), *Plynlimon Research: The First Two Decades*, Report 109 (Institute of Hydrology: Wallingford, UK).

Kittredge, J. (1948/1973), *Forest Influences* (Dover: New York).

Lees, M. L. (1985), *Inland water surveying in the UK: A short history*, Hydrological Data UK 1985 Yearbook (Institute of Hydrology: Wallingford, UK).

L'vovich, M. I., White, G. F., and collaborators (1990), 'Use and transformation of terrestrial water systems', in B. L. Turner *et al.* (eds.), *The Earth as Transformed by Human Action* (Cambridge), 235–52.

Marsh, G. P. (1864/1965), *Man and Nature: Physical Geography as Modified by Human Action* (repr.), (Belknap Press of Harvard University Press: Cambridge, Mass.).

Marsh, T., and Lees, M. (1985), *The 1984 Drought*, Hydrological Data UK (Institute of Hydrology/British Geological Survey: Wallingford, UK).

Meyer-Homji, V. M. (1988), 'Effects of forests on precipitation in India', in E. R. C. Reynolds, and F. B. Thompson (eds.), *Forests, Climate and Hydrology: Regional Impacts* (United Nations University: Tokyo), 51–77.

Ministry of Agriculture, Fisheries, and Food (1967), *Potential Transpiration* (HMSO: London).

Monteith, J. L. (1965), 'Evaporation and environment', *Symposium, Society of Expermental Biologists*, 19: 205–34.

—— and Unsworth, M. H. (1990), *Principles of Environmental Physics*, 2nd edn. (Edward Arnold: London).

Newson, M. D. (1992), *Land, Water and Development: River Basin Systems and their Management* (Routledge: London).

Oyebande, L. (1988), 'Effects of tropical forest on water yield', in E. R. C. Reynolds, and F. B. Thompson (eds.), *Forests, Climate and Hydrology: Regional Impacts* (United Nations University: Tokyo), 16–50.

Penman, H. L. (1948), 'Natural evaporation from open water, bare soil and grass', *Proceedings of the Royal Society of London*, Ser. A, 193: 120–45.

Reynolds, E. R. C., and Thompson, F. B. (eds.) (1988), *Forests, Climate and Hydrology: Regional Impacts* (United Nations University: Tokyo).

Richards, K. S., Brunsden, D., Jones, D. K. C., and McCaig, M. (1987), 'Applied fluvial geomorphology: River engineering project appraisal in its geomorphological context, in K. S. Richards (ed.), *River Channels: Environment and Process* (Blackwell: Oxford), 348–82.

Rodda, J. C., and Smith, S. W. (1986), 'The significance of the systematic error in rainfall measurement for assessing wet deposition', *Atmospheric Environment*, 20/5: 1059–64.

Shaw, E. M. (1989), *Engineering Hydrology Techniques in Practice* (Ellis Horwood: Chichester, UK).

Shiklomanov, I. A., and Krestovsky, O. I. (1988), 'The influence of forests and forest reclamation practice on streamflow and water balance', in E. R. C. Reynolds, and F. B. Thompson (eds.), *Forests, Climate and Hydrology: Regional Impacts* (United Nations University: Tokyo), 78–116.

Smart, J. D. C. (1977), *The design, operation and calibration of the permanent flow measuring structures in the Plynlimon expermental catchments*, Report 42 (Institute of Hydrology: Wallingford, UK).

Soil Survey of England and Wales (1974*a*), *Soil Survey Field Handbook* (Harpenden, UK).

—— (1974*b*), *Soil Survey Laboratory Methods* (Harpenden, UK).

Strahler, A. H., and **Strahler, A. N.** (1992), *Modern Physical Hydrology*, 4th edn. (John Wiley & Sons: New York).

Wilson, E. M. (1983), *Engineering Hydrology* (3rd edn.) (Macmillan: Basingstoke, UK).

Winter, E. J. (1974), *Water, Soil and the Plant* (Macmillan: Basingstoke,, UK).

World Resources Institute (1990), *World Resources 1990–91* (Oxford).

Chapter 3

Anderson, M. G., and **Burt, T. P.** (1978), 'The role of topography in controlling throughflow generation', *Journal of Hydrology* 33: 383–90.

—— —— (eds.) (1985), *Hydrological Forecasting* (John Wiley & Sons: Chichester, UK).

Atkinson, T. C. (1978), 'Techniques for measuring subsurface flow on hillslopes', in M. J. Kirkby (ed.), *Hillslope Hydrology* (John Wiley & Sons: Chichester, UK), 73–120.

Beven, K., Gilman, K., and **Newson, M. D.** (1979), 'Flow and flow routeing in upland channel networks', *Hydrological Sciences Bulletin*, 24/3: 303–25.

Burch, G. J., Moore, I. D., and **Burns, J.** (1989), 'Soil hydrophobic effects on infiltration and catchment runoff', *Hydrological Processes*, 3/3: 211–22.

Buttle, J. M. (1990), 'Effects of suburbanization upon snowmelt runoff', *Hydrological Sciences Journal*, 35/3: 285–302.

Dunne, T., and **Leopold, L. B.** (1978), *Water in Environmental Planning* (Freeman: San Francisco).

Ferguson, R. I. (1984), 'Magnitude and modelling of snowmelt runoff in the Cairngorm Mountains, Scotland', *Hydrological Sciences Journal*, 29/1: 49–62.

Ford, D. C., and **Williams, P. W.** (1989), *Karst Geomorphology and Hydrology* (Unwin Hyman: London).

Gee, G. W., and **Hillel, D.** (1988), 'Groundwater recharge in arid regions: Review and critique of estimation methods', *Hydrological Processes*, 2: 255–66.

Gilman, K., and **Newson, M. D.** (1980), *Soil Pipes and Pipeflow*, British Geomorphological Research Group, Research Monograph 1 (Geo Books: Norwich, UK).

Gurnell, A. M., Hughes, P. A., and **Edwards, P. J.** (1990), 'The hydrological implications of heath vegetation composition and management in the New Forest, Hampshire, England', in J. B. Thornes (ed.), *Vegetation and Erosion* (John Wiley & Sons: Chichester, UK), 179–98.

Hewlett, J. D. (1982), *Principles of Forest Hydrology* (University of Georgia Press: Athens, Ga.).

Hollis, G. E., and **Ovenden, J. C.** (1988), 'One year irrigation experiment to assess losses and runoff volume relationships for a residential road in Hertfordshire, England', *Hydrological Processes*, 2: 61–74.

Horton, R. E. (1945), 'Erosional development of streams and their drainage basins: Hydrological approach to quantitative morphology', *Geological Society of America Bulletin*, 56: 275–370.

Ingram, H. A. P., Rycroft, D. W., and **Williams, D. J. A.** (1974), 'Anomalous transmission of water through certain peats', *Journal of Hydrology*, 22: 213–18.

Institute of Hydrology (1980), *Low Flood Studies* (Wallingford, UK).

Jackson, I. J. (1989), *Climate, Water and Agriculture in the Tropics* (2 edn.) (Longman: Harlow, UK).

Jones, J. A. A. (1981), 'The nature of soil piping: A review of research', *BGRG Research Monograph*, 3.

Kirby, C. (1984), *Water in Great Britain* (Penguin Books: Harmondsworth, UK).

Kirkby, M. J. (ed.) (1978), *Hillslope Hydrology* (John Wiley & Sons: Chichester, UK).

Ledger, D. C. (1981), 'The velocity of the River Tweed and its tributaries', *Freshwater Biology*, 11: 1–10.

Leopold, L. B., Wolman, M. G., and **Miller, J. P.** (1964), *Fluvial Processes in Geomorphology* (W. H. Freeman: San Francisco).

Lewin, J. (1989), 'Floods in fluvial geomorphology', in K. Beven and P. Carling (eds.), *Floods: Hydrological, Sedimentological and Geomorphological Implications* (John Wiley & Sons: Chichester, UK), 265–84.

Lloyd, J. W. (1986), 'A review of aridity and groundwater', *Hydrological Processes*, 1: 63–78.

Mahoo, H. (1989), 'Deforestation of a tropical humid rainforest and resulting effects on soil properties, surface and subsurface flow, water quality and crop evapotranspiration', unpublished PhD thesis, University of Sokoine, Morogoro, Tanzania.

Ministry of Agriculture, Fisheries and Food (1982), *The Design of Field Drainage Pipe Systems*, Reference Book 345 (HMSO: London).

Morris, E. M. (1985), 'Snow and Ice', in M. G. Anderson, and T. P. Burt (eds.), *Hydrological Forecasting* (John Wiley & Sons: Chichester, UK), 153–82.

Newson, M. D. (1979), *The Physiography, Deposits and Vegetation of the Plynlimon Catchments*, Report 30 (Institute of Hydrology: Wallingford, UK).

Pilgrim, D. H. (1983), 'Some problems in transferring hydrological relationships between small and large drainage basins and between regions', *Journal of Hydrology*, 65: 49–72.

—— **Chapman, T. G.,** and **Doran, D. G.** (1988), 'Problems of rainfall-runoff modelling in arid and semi-arid regions', *Hydrological Sciences Journal*, 33/4: 379–400.

Price, M. (1985), *Introducing Groundwater* (George Allen & Unwin: London).

Reid, I., and **Frostick, L. E.** (1987), 'Flow dynamics and suspended sediment properties in arid zone flash floods', *Hydrological Processes*, 1/3: 239–53.

Reisner, M. (1990), *Cadillac Desert* (Secker & Warburg: London).

Richards, K. (1982), *Rivers, Form and Process in Alluvial Channels* (Methuen: London).

Rivett, M. O., Lerner, D. N., and **Lloyd, J. W.** (1990), 'Chlorinated solvents in UK aquifers', *Journal of the Institution of Water and Environmental Management*, 4/3: 242–50.

Rowley, G. (1990), 'The West Bank: Native water-resource systems and competition', *Political Geography Quarterly*, 9/1: 39–52.

Smedema, L. K., and **Rycroft, D. W.** (1983), *Land Drainage: Planning and Design of Agricultural Drainage Systems* (Batsford: London).

Strahler, A. (1957), 'Quantitative analysis of watershed geomorphology', *Transactions of the American Geophysical Union*, 38: 913–20.

Warner, R. F. (1987), 'Spatial adjustments to temporal variations in flood regime in some Australian rivers', in K. Richards (ed.), *River Channels: Environment and Process* (Blackwell, Oxford), 14–40.

Weyman, D. R. (1975), *Runoff Processes and Streamflow Modelling* (Oxford).

Wright, J. F., and **Berrie, A. D.** (1987), 'Ecological effects of groundwater pumping and a natural drought on the upper reaches of a chalk stream', *Regulated Rivers: Research and Management*, 1: 145–60.

Yair, A., and Lavee, H. (1985), 'Runoff generation in arid and semi-arid zones', in M. G. Anderson and T. P. Burt (eds.), *Hydrological Forecasting* (John Wiley & Sons: Chichester, UK), 183–220.

Chapter 4

Acreman, M. C. (1989), 'Extreme rainfall in Calderdale, 19th May, 1989', *Weather*, 44/11: 438–46.

Agnew, C., and Anderson, E. (1992), *Water Resources in the Arid Realm* (Routledge: London).

American Society of Civil Engineers (1975, 1988), *Lessons from Dam Incidents, USA*, i and ii. (New York).

Anderson, M. G., and Burt, T. P. (eds.) (1985), *Hydrological Forecasting* (John Wiley & Sons: Chichester, UK).

Beaumont, P. (1989), *Environmental Management and Development in Drylands* (Routledge: London).

Benson, M. A. (1962), *Factors Influencing the Occurrence of Floods in a Humid Region of Diverse Terrain*, Professional Paper, 1580B, United States Geological Survey.

Beven, K., and Carling, P. (eds.) (1989), *Floods, Hydrological, Sedimentological and Geomorphological Implications* (John Wiley & Sons: Chichester, UK).

Brammer, H. (1990a) 'Floods in Bangladesh: Geographical background to the 1987 and 1988 floods', *Geographical Journal*, 156/1: 12–22.

—— (1990b) 'Floods in Bangladesh: Flood mitigation and environmental aspects', *Geographical Journal*, 156/2: 158–65.

Brown, J. (ed.) (1989), *Environmental Threats: Perception, Analysis and Management* (Belhaven: London).

Burton, I., Kates, R. W., and White, G. F. (1978), *The Environment as Hazard* (Oxford University Press, New York).

Church, M. (1988), 'Floods in cold climates', in V. R. Baker, R. C. Kochel, and P. C. Patton (eds.), *Flood Geomorphology* (John Wiley & Sons: New York), 205–29.

Clark, C. (1983), *Flood* (Time-Life: Amsterdam).

Clarke, R. (1991), *Water: The International Crisis* (Earthscan: London).

Collinge, V., and Kirby, C. (eds.) (1987), *Weather Radar and Flood Forecasting* (John Wiley & Sons: Chichester, UK).

Cooke, R. U., Brunsden, D., Doornkamp, J. C., and Jones, D. K. C. (1982), *Urban Geomorphology in Drylands* (Oxford).

Costa, J. E. (1988), 'Floods from dam failures', in V. R. Baker, R. C. Kochel, and P. C. Patton (eds.), *Flood Geomorphology* (John Wiley & Sons: New York), 439–63.

Freedman, W. (1989), *Environmental Ecology: The Impacts of Pollution and other Stresses on Ecosystem Structure and Function* (Academic Press: London).

Gardiner, J. L. (ed.) (1991), *River Projects and Conservation: A Manual for Holistic Appraisal* ((John Wiley & Sons: Chichester, UK).

Graf, W. L. (1985), *The Colorado River: Instability and Basin Management* (Association of American Geographers: Washington, DC).

Gray, D. M. (1964), 'Physiographic characteristics and the runoff pattern', *Journal of Geophysical Research*, 66: 1215–23.

Gupta, A. (1988), 'Large floods as geomorphic events in the humid tropics', in V. R. Baker, R. C. Kochel, and P. C. Patton (eds.), *Flood Geomorphology* (John Wiley & Sons: New York), 301–15.

Gupta, S. N. (1969), *Statistical correlation of Himalayan and Bundelkhand basin characteristics with flood flows* Publication 85 (International Association of Scientific Hydrologists), 874–80.

Gustard, A., Cole, G., Marshall, D., and Bayliss, A. (1987), *A Study of Compensation Flows in the UK* Report 99 (Institute of Hydrology: Wallingford, UK).

Hamlin, M. J., and Wright, C. E. (1978), 'The effects of drought on the river system', *Proceedings of the Royal Society of London*, Series A, 363: 69–96.

Hanwell, J. D., and Newson, M. D. (1970), *The Great Storms and Floods of July 1968 on Mendip*, Wessex Cave Club Occasional Publication, 1/2 (Priddy: Somerset, UK).

Hayden, B. P. (1988), 'Flood climates', in V. R. Baker, R. C. Kochel, and P. C. Patton (eds.), *Flood Geomorphology* (John Wiley & Sons: New York), 13–26.

Hewitt, K., and Burton, I. (1971), *The Hazardousness of a Place: A Regional Ecology of Damaging Events* (University of Toronto Press).

Higgins, G. M., Dielman, P. J., and Abernethy, C. L. (1988), 'Trends in irrigation development and their implications for hydrologists and water resources engineers', *Hydrological Sciences Journal*, 33/1 and 2, 43–59.

Hirschboeck, K. K. (1988), 'Flood hydroclimatology', in V. R. Baker, R. C. Kochel, and P. C. Patton (eds.), *Flood Geomorphology* (John Wiley & Sons: New York), 27–49.

Jansson, K., Harris, M., and Penrose, A. (1987), *The Ethiopian Famine* (Zed Books: London).

Johnson, P. (1975) 'Snowmelt'. Paper presented at Flood Studies Conference (Institute of Civil Engineers, London), 5–10.

Kates, R. W. (1978), *Risk Assessment of Environmental Hazard*, SCOPE 8 (John Wiley & Sons: Chichester, UK).

Leese, M. D. (1973), 'Use of censored data in the estimation of Grumbel distribution parameters for annual maximum flood series', *Water Resources Research*, 9: 1534–42.

Lewis, G., and Williams, G. (1984), *Rivers and Wildlife Handbook: A Guide to Practices which further the Conservation of Wildlife on Rivers* (Royal Society for the Protection of Birds: Sandy, UK).

McCahon, C. P., Carling, P. A., and Pascoe, D. (1987), 'Chemical and ecological effects of a Pennine peat-slide, *Environmental Pollution*, 45: 275–89.

McEwen, L. (1989), 'River channel changes in response to flooding in the Upper River Dee catchment, Aberdeenshire over the last 200 years', in K. Beven, and P. Carling (eds.), *Floods: Hydrological, Sedimentological and Geomorphological Implications* (John Wiley & Sons: Chichester, UK), 219–38.

Macklin, M. G., Rumsby, B. T., and Newson, M. D. (1992), 'Historic overbank floods and floodplain sedimentation in the lower Tyne Valley, North-East England', in R. D. Hey (ed.), *Gravel-bed Rivers* (John Wiley & Sons: Chichester, UK).

McMahon, T. A., Finlayson, B. L., Haines, A., and Srikanthan, R. (1987), 'Runoff variability: A global perspective', in S. I. Solomon, M. A. Beran, and W. Hogg (eds.), *The Influence of Climate Change and Climate Variability on the Hydrologic Regime and Water Reesources* Publication 168 (International Association of Hydrological Sciences), 3–11.

Milliman, J. D., Broadus, J. M., and Gable, F. (1989), 'Environmental and economic implications of rising sea-level and subsiding deltas: The Nile and Bengal examples', *Ambio*, 6: 340–5.

Ministry of Agriculture, Fisheries, and Food (1982), *Irrigation*, Reference Book 138 (HMSO: London).

Natural Environmental Research Council (NERC) (1975), *Flood Studies Report* (5 vols).

Nash, J. E. (1960), 'A unit hydrograph study with particular reference to British catchments', *Proceedings of the Institution of Civil Engineers*, 17: 249–82.

Newson, M. D., and Lewin, J. (1991), 'Climatic change, river flow extremes and fluvial erosion: Scenarios for England and Wales', *Progress in Physical Geography*, 15/1: 1–17.

Nixon, M. (1966), 'Flood regulation and river training', in R. B. Thorn (ed.), *River Engineering and Water Conservation Works* (Butterworth: London).

Penning-Rowsell, E. C., and Chatterton, J. B. (1977), *The Benefits of Flood Alleviation: A Manual of Assessment Techniques* (Saxon House: Farnborough, UK).

Petts, G. E. (1984), *Impounded Rivers* (John Wiley & Sons: Chichester, UK).

Richards, K. S., Brunsden, D., Jones, D. K. C., and McCaig, M. (1987), 'Applied fluvial geomorphology: River engineering project appraisal in its geomorphological context', in K. S. Richards (ed.), *River Channels, Environment and Process* (Blackwell: Oxford), 48–32.

Rodda, J. C. (1967), *The significance of characteristics of basin rainfall and morphometry in a study of floods in the United Kingdom*, Publication 85 (International Association of Scientific Hydrologists), 834–45.

Rodier, J., and Roche, M. (1978), 'River flow in arid regions', in R. W. Herschey (ed.), *Hydrometry* (John Wiley & Sons: Chichester, UK), 453–72.

Schick, A. (1988), 'Hydrologic aspects of floods in extreme arid environments', in V. R. Baker, R. C. Kochel, and P. C. Patton (eds.), *Flood Geomorphology* (John Wiley & Sons, New York), 189–203.

Shaw, E. M. (1983), *Hydrology in Practice* (Van Nostrand Reinhold, UK: Wokingham).

Sherman, L. K. (1932), 'Streamflow from rainfall by the unit-graph method', *Engineering News Record*, 108: 501–5.

Smith, N. (1972), *A History of Dams* (Citadel Press: Secaucus, NJ).

Sutcliffe, J. V., Dugdale, G., and Milford, J. R. (1989), 'The Sudan floods of 1988', *Hydrological Sciences Journal*, 34/3, 355–64.

UNEP (1991), *United Nations Environment Programme Environmental Data Report* (Blackwell: Oxford).

White, G. F. (1964), *Choice of Adjustments to Floods*, Department of Geography Research Paper 93 (University of Chicago Press).

Wijkman, A., and Timberlake, L. (1984), *Natural Disasters: Acts of God or Acts of Man?* (Earthscan: London).

Wilson, E. M. (1983), *Engineering Hydrology* (3rd edn.) (Macmillan: Basingstoke, UK).

World Commission on Environment and Development (1986), *Our Common Future* (Oxford).

Chapter 5

Acreman (1985) 'The effects of afforestation on the flood hydrology of the upper Ettrick Valley', *Scottish Geographical Magazine*, 39/2: 89–98.

Alvares, C., and Billorey, R. (1988), *Damming the Narmada: India's Greatest Planned Environmental Disaster* (Third World Network/Asia-Pacific People's Environment Network: Malaysia).

Arnell, N. (1989), *Human Influences on Hydrological Behaviour: An International Literature Survey* (UNESCO: Paris).

Atkinson, B. (1979), 'Urban influences on precipitation in London', in G. E. Hollis (ed.), *Man's Influence on the Hydrological Cycle in the United Kingdom* (Geo Books: Norwich), 123–33.

Boardman, J. (1988), 'Public policy and soil erosion in Britain', in J. M. Hooke (ed.), *Geomorphology in Environmental Planning* (John Wiley & Sons, Chichester, UK), 33–50.

Bosch, J. M., and Hewlett, J. D. (1982), 'A review of catchment experiments to determine the effects of vegetation changes on water yield and evapotranspiration', *Journal of Hydrology*, 55: 3–23.

Boucher, K. (1990), 'Landscape and technology, the Gabcikovo–Nagymaros scheme', in D. Cosgrove, and G. Petts (eds.), *Water, Engineering and Landscape* (Belhaven: London), 174–87.

Brookes, A. (1988), *Channelized Rivers: Perspectives for Environmental Management* (John Wiley & Sons: Chichester, UK).

Bruenig, E. F. (1986), 'Aspects of current forestry practice and silvicultural trends in West Germany affecting fresh waters', in J. F. de L. G. Solbe (ed.), *Effects of Land Use on Fresh Waters* (Ellis Horwood: Chichester, UK), 378–97.

Calder, I. R. (1980), *The Effects of Afforestation on Runoff from the Catchments Supplying Crinan Canal Reservoirs*, Report to British Waterways Board (Institute of Hydrology: Wallingford).

—— (1990), *Evaporation in the Uplands* (John Wiley & Sons: Chichester, UK).

—— and Newson, M. D. (1979), 'Land use and upland water resources in Britain: A strategic look', *Water Resources Bulletin*, 15/6. 1628–39.

—— —— and Walsh, P. D. (1982), 'The application of catchment, lysimeter and hydro-meteorological studies to land-use planning and water management', Special Edition of Proceedings of the International Symposium held in Berne, September 1982, *Hydrological Research Basins and their Use in Water Resources Planning* (Sonderh. Landeshyd.: Berne), 3: 853–63.

—— Wright, I. R., and Murdiyarso, D. (1986), 'A study of evaporation from tropical rain-forest, West Java', *Journal of Hydrology*, 89: 13–33.

Cummings, B. J. (1990), *Dam the Rivers, Damn the People: Development and Resistance in Amazonian Brazil* (Earthscan: London).

Ellis, J. B. (1979), 'The nature and sources of urban sediments and their relation to water quality: A case study from north-west London', in G. E. Hollis (ed.), *Man's Impact on the Hydrological Cycle in the United Kingdom* (Geo Books: Norwich), 199–216.

Essery, C. I., and Wilcock, D. N. (1990), 'The impact of channelization on the hydrology of the Upper River Man, County Antrim, Northern Ireland: A long term case study', *Regulated Rivers: Research and Management*, 5: 17–34.

Flader, S. (1987), 'Aldo Leopold and the evolution of a land ethic', in T. Tanner (ed.), *Aldo Leopold, the Man and his Legacy* (Soil Conservation Society of America: Ankeny, Ia), 3–24.

Forestry Commission (1988), *Forestry and Water Guidelines* (Forestry Commission: Edinburgh).

Gash, J. H. C., Wright, I. R., and Lloyd, C. R. (1980), 'Comparative estimates of interception loss from three coniferous forests in Great Britain', *Journal of Hydrology*, 48: 89–105.

George, R. J. (1990), 'Reclaiming sandplain seeps by intercepting perched groundwater with Eucalypts', *Land Degradation and Rehabilitation*, 2: 13–25.

Gore, J. A. (1985), *The Restoration of Rivers and Streams: Theories and Experience* (Butterworth: Boston).

Graf, W. L. (1985), *The Colorado River: Instability and Basin Management* (Association of American Geographers: Washington, DC).

Greenwood, E. A. N., Klein, L., Beresford, J. D., and Watson, G. D. (1985), 'Differences in annual evaporation between grazed pasture and Eucalyptus species in plantations on a saline farm catchment', *Journal of Hydrology*, 78: 261–78.

Gregoire, J. M. (1991), 'Watershed degradation: Use of thermal data and vegetation indices as indicators of environmental changes: Hydrological implications of changes in land surface cover', in A. S. Belward, and C. R. Valenzuela (eds.), *Remote Sensing and*

GIS for Resource Management in Developing Countries (Kluwer: Dordrecht, Neths.), 235–52.

Higgins, G. M., Dielman, P. J., and Abernethy, C. L. (1988), 'Trends in irrigation development and their implications for hydrologists and water resource engineers', *Hydrological Sciences Journal*, 33/1/2: 43–59.

Higgs, G. (1987), 'Environmental change and hydrological response: Flooding in the Upper Severn Catchment', in K. J. Gregory, J. Lewin, and J. B. Thornes (eds.), *Palaeohydrology in Practice* (John Wiley & Sons: Chichester, UK), 131–59.

Hollis, G. E. (1974), 'The effect of urbanisation on floods in the Canon's Brook, Harlow, Essex', in K. J. Gregory, and D. E. Walling (eds.), *Fluvial Processes in Instrumented Watersheds*, Special Publication 6 (Institute of British Geographers), 123–39.

—— (1975), 'The effect of urbanisation on floods of different recurrence intervals', *Water Resources Research*, 11: 431–4.

—— (ed.) (1979), *Man's Impact on the Hydrological Cycle in the United Kingdom* (Geo Books: Norwich).

—— and Ovenden, J. C. (1988a), 'One year irrigation experiment to assess losses and runoff volume relationships for a residential road in Hertfordshire, England', *Hydrological Processes*, 2: 61–74.

—— —— (1988b), 'The quantity of stormwater runoff from ten stretches of road, a car park and eight roofs in Hertfordshire, England, during 1983', *Hydrological Processes*, 2: 227–43.

Kirby, C., Newson, M. D., and Gilman, G. (1991), *Plynlimon Research: The First Two Decades*, Report 109 (Institute of Hydrology: Wallingford, UK).

Knight, C. (1979), 'Urbanization and natural stream channel morphology: The case of two English new towns', in G. E. Hollis (ed.), *Man's Impact on the Hydrological Cycle in the United Kingdom* (Geo Books: Norwich), 181–98.

Landsberg, D. (1960) *Physical Climatology* (Gray Printing Co.: Dubois, Pa.).

Law, F. (1956), 'The effect of afforestation upon the yield of water catchment areas', *Journal of the British Waterworks Association*, 38: 484–94.

Leopold, L. B. (1968), *Hydrology for Urban Land Planning: A Guidebook on the Hydrologic Effects of Urban Land Use*, Circular 554 (United States Geological Survey).

—— (1991), 'Lag times for small drainage basins', *Catena*, 18: 157–71.

Lindh, G. (1990), 'Functions and uses of water in urbanized areas', in H. Massing, J. Packman, and F. C. Zuidema (eds.), *Hydrological Processes and Water Management in Urban Areas*, Publication 198 (International Association of Hydrological Sciences), 71–82.

L'vovitch, M. I., and White, G. F. (1990), 'Use and transformation of terrestrial water systems', in B. L. Turner *et al.* (eds.), *The Earth as Transformed by Human Action* (Cambridge), 235–52.

Macklin, M. G., and Lewin, J. (forthcoming), 'Holocene river alluviation in Britain', in I. Douglas, and J. Hagedorn, *Zeitschrift fur Geormophologie*, (Suppl. 85).

Maitland, P. S., Newson, M. D., and Best, G. A. (1990), 'The impact of afforestation and forestry practice on freshwater habitats', *Focus on Conservation*, 23 (Nature Conservancy Council, UK).

Moffat, A. J. (1989), 'Forestry and soil erosion in Britain: A reply', *Soil Use and Management*, 5: 199–200.

Newson, M. D. (1991), 'Catchment control and planning: Emerging patterns of definition, policy and legislation in UK water management', *Land Use Policy*, 8/1: 9–15.

—— (1992), *Land, Water and Development* (Routledge: London).

Newson, M. D. (1989), 'Forests and water resources: Problems of prediction on a regional scale', *Philosophical Transactions of the Royal Society, London*, B324, 283–98.

—— and **Robinson, M.** (1983), 'Effects of agricultural drainage on upland streamflow: Case studies in mid-Wales', *Journal of Environmental Management*, 17: 333–48.

OECD (1989), *Waer Resource Management: Integrated Policies* (OECD: Paris).

O'Riordan, J. (1986), 'Some examples of land- and water-use planning in British Columbia', in F. T. Last, M. C. B. Hotz, and B. G. Bell (eds.), *Land and its Uses, Actual and Potential* (Plenum Press: New York), 193–211.

Oyebande, L. (1988), 'Effects of tropical forest on water yield', in E. R. C. Reynolds and F. B. Thompson (eds.), *Forests, Climate and Hydrology: Regional Impacts* (United Nations University: Tokyo), 16–50.

Packman, J. C. (1979), 'The effect of urbanization on flood magnitude and frequency', in G. E. Hollis (ed.), *Man's Impact on the Hydrological Cycle in the United Kingdom* (Geo Books: Norwich, UK).

Park, C. C. (1977), 'Man-induced changes in stream channel capacity', in K. J. Gregory (ed.), *River Channel Changes* (John Wiley & Sons: Chichester, UK), 121–44.

Penman, H. L. (1963), *Vegetation and Hydrology* (Commonwealth Bureau of Soils: Harpenden, UK).

Petts, G. E. (1979), 'Complex response of river channel morphology subsequent to reservoir construction', *Progress in Physical Geography*, 3/3: 329–62.

—— (1984), *Impounded Rivers: Perspectives for Ecological Management* (John Wiley & Sons: Chichester, UK).

—— (1987), 'Timescales for ecological change in regulated rivers', in J. Craig, and J. B. Kemper (eds.), *Regulated Streams: Advances in Ecology* (Plenum Press: New York), 257–66.

Pipe, B. S., Gustard, A., Green, C. S., and **Pachern Sridurongkatum** (1991), 'Water resource developments and flow regimes on the Mekong River', in F. H. M. Van de Ven, D. Gutknecht, D. P. Loucks, and K. A. Salewicz (eds.), *Hydrology for the Water Management of Large River Basins*, Publication 201 (International Association of Hydrological Sciences) 45–56.

Ponce, S. L. (1986), 'Controlling diffuse-source pollution association with forest practices in North America', in J. F. de L. G. Solbe (ed.), *Effects of Land Use on Fresh Waters* (Ellis Horwood: Chichester, UK), 432–43.

Province of Ontario (1987), *Urban Drainage Design Guidelines* (Government Printing Office: Toronto).

Reid, I. (1979), 'Seasonal changes in microtopography and surface depression storage of arable soils', in G. E. Hollis (ed.), *Man's Impact on the Hydrological Cycle in the United Kingdom* (Geo Books: Norwich), 19–30.

—— and **Parkinson, R. J.** (1984), 'The nature of the tile drain outfall hydrograph in heavy clay soils', *Journal of Hydrology*, 72: 289–305.

Reynolds, S. G. (1990), 'The influence of forest-clearance methods, fellage and slope runoff on soil chemical properties and banana plant yields in the South Pacific', in J. Boardman, I. D. L. Foster, and J. A. Dearing (eds.), *Soil Erosion on Agricultural Land* (John Wiley & Sons: Chichester, UK), 339–50.

Roberts, C. (1989), 'Flood frequency and urban-induced channel change: Some British examples', in K. Beven, and P. Carling (eds.), *Floods; Hydrological, Sedimentological and Geomorphological Implications* (John Wiley & Sons: Chichester, UK), 57–82.

Roberts, J. M. (1983), 'Forest transpiration: A conservative hydrological process?' *Journal of Hydrology*, 66: 133–41.

Robinson, M. (1986), 'Changes in catchment runoff following drainage and afforestation', *Journal of Hydrology*, 86: 71–84.

—— and Armstrong, A. C. (1988), 'The extent of agricultural field drainage in England and Wales', *Transactions, Institute of British Geographers*, 13: 19–28.

—— and Beven, K. J. (1983), 'The effect of mole drainage on the hydrological response of a swelling clay soil', *Journal of Hydrology*, 63: 205–23.

—— and Newson, M. D. (1986), 'Comparison of forest and moorland hydrology in an upland area with peat', *International Peat Journal*, 1/1: 49–68.

Robinson, M. D. (1990), *Impact of Improved Land Drainage on River Flows*, Report 113 (Institute of Hydrology: Wallingford, UK).

Ross, S. M., Thorne, J. B., and Nortcliffe, S. (1990), 'Soil hydrology, nutrient and erosional response to the clearance of terra firme forest, Maraca Island, Roraima, northern Brazil', *Geographical Journal*, 156/3: 267–82.

Schwartz, H. E., Emel, J., Dickens, W. J., Rogers, P., and Thompson, J. (1991), 'Water quality and flows', in B. L. Turner *et al.* (eds.), *The Earth as Transformed by Human Action* (Cambridge), 253–70.

Shaw, E. M. (1983), *Hydrology in Practice* (Van Nostrand Reinhold, UK: Wokingham).

Shuttleworth, W. J. (1991), 'Insight from large-scale observational studies of land/atmosphere interactions', *Surveys in Geophysics*, 12: 3–30.

Solbe, J. F. de L. G. (ed.) (1986), *Effects of Land Use on Fresh Waters: Agriculture, Forestry, Mineral Exploitation, Urbanisation* (Ellis Horwood: Chichester, UK).

Soutar, R. G. (1989), 'Afforestation and sediment yields in British fresh waters', *Soil Use and Management*, 5: 82–6.

Speirs, R. B., and Frost, C. A. (1985), 'The increased incidence of accelerated soil water erosion on arable land in the east of Scotland', *Research and Development in Agriculture*, 2: 161–7.

Swanson, R. H. (1973), 'Small openings in poplar forest increase snow accumulation', *UNESCO/WMO/IAHS, Banff Symposium*, ii. 1382–9.

Szolgay, J. (1991), 'Prediction of river runoff changes due to hydropower development on the Danube at Gabcikovo', in F. H. M. Van de Ven, D. Gutknecht, D. P. Louks, and K. A. Salewicz (eds.), *Hydrology for the Water Management of Large River Basins*, Publication 201 (IAHS), 209–18.

UNEP/UNESCO (1990), *The Impact of Large Water Projects on the Environment* (UNESCO: Paris).

Ven, F. H. M. van de, Gutknecht, D., Loucks, D. P., and Salewicz, K. A. (eds.) (1991), *Hydrology for the Water Management of Large River Basins*, Publication 201 (IAHS).

Veneklaas, E. J., and Ek, R. van (1990), 'Rainfall interception in two tropical montane rain forests, Colombia', *Hydrological Processes*, 4: 311–26.

Williams, W. D., and Aladin, N. V. (1991), 'The Aral Sea: Recent limnological changes and their conservation significance', *Aquatic Conservation: Marine and Freshwater Ecosystems*, 1: 3–23.

Woolhouse, C. H. (1989), 'Managing the effects of urbanisation in the Upper Lee', *Second National Hydrology Symposium, British Hydrological Society*, 2.9–2.17.

Chapter 6

Schumm, S. A. (1977), *The Fluvial System* (John Wiley & Sons: New York).

Chapter 7

Advisory Centre for Education (1971), *Clean Water: The Water of Rivers and Streams, the Life it Supports and the Effects of Pollution*, Things of Science kit (Cambridge).

Baldock, D. (1984), *Wetland Drainage in Europe: The Effects of Agricultural Policy in Four EEC Countries* (IEEP/IIED: London).

Bjork, S., and **Digerfeldt, G.** (1991), 'Development and degradation, redevelopment and preservation of Jamaican wetlands', *Ambio*, 20/7: 276–84.

Boon, P. J. (1991), 'Environmental Impact Assessment and the water industry: Implications for nature conservation', *Journal of the Institution of Water and Environmental Management*, 5: 194–205.

British Ecological Society (1990), *River Water Quality*, Environmental Issues, 1 (Field Studies Council: London).

Brookes, A. (1988), *Channelized Rivers: Perspectives for Environmental Management* (John Wiley & Sons: Chichester, UK).

—— **Gregory, K. J.,** and **Dawson, F. H.** (1983), 'An assessment of river channelization in England and Wales', *Science of the total environment*, 27: 97–112.

Burgis, M. J., and **Morris, P.** (1987), *The Natural History of Lakes* (Cambridge).

Burton, A. (1982), *The Changing River: An Account of some British Rivers and their History* (Victor Gollancz: London).

Canter, L. (1985), *Environmental Impact of Water Resources Projects* (Lewis Publishers: Chelsea, Mich.).

Castle, D. A., McCunnall, J., and **Tring, I. M.** (1984), *Field Drainage: Principles and Practices* (Batsford: London).

Darby, H. C. (1983), *The Changing Fenland* (Cambridge).

Extence, C. A., and **Ferguson, A. J. D.** (1989) 'Aquatic invertebrate surveys as a water management tool in the Anglian Water region', *Regulated Rivers: Research and Management*, 4/2: 139–46.

Freedman, B. (1989), *Environmental Ecology* (Academic Press: San Diego, Calif.).

Gilman, K., and **Newson, M. D.** (1982), *The Anglesey Wetlands Study*, Report to Nature Conservancy Council (Institute of Hydrology: Wallingford, UK).

Gore, J. A., and **Nestler, J. M.** (1988), 'Instream flow studies in perspective', *Regulated Rivers: Research and Management*, 2: 93–101.

—— —— and **Layzer, J. B.** (1989), 'Instream flow predictions and management options for biota affected by peaking-power hydroelectric operations', *Regulated Rivers: Research and Management*, 3: 35–48.

Haslam, S. M. (1978), *River Plants. The Macrophytic Vegetation of Watercourses* (Cambridge).

—— and **Wolseley, P. A.** (1981), *River Vegetation: Its Identification, Assessment and Management* (Cambridge).

Hellawell, J. M. (1986), *Biological Indicators of Freshwater Pollution and Environmental Management* (Elsevier: London).

Hollis, G. E. (1990), 'Environmental impacts of development on wetlands in arid and semi-arid lands', *Hydrological Sciences Journal*, 35/4: 411–28.

Holmes, N. T. H. (1983), *Typing British Rivers according to their Flora* (Nature Conservancy Council, Focus on Conservation: London).

—— **Driver, A.,** and **Brookes, A.** 'Wildlife', in J. A. Gardiner (ed.), *River Projects and Conservation: A Manual for Holistic Appraisal* (John Wiley & Sons: Chichester, UK), 169–84.

Ingram, H. A. P. (1982) 'Size and shape in raised mire ecosystems: A geophysical model', *Nature*, 297 (5864): 300–3.

Jeffries, M., and **Mills, D.** (1990), *Freshwater Ecology: Principles and Applications* (Belhaven Press: London).

Johnson, P. (1988), 'River regulation: A regional perspective, Northumbrian Water Authority', *Regulated Rivers: Research and Management*, 2: 233–55.

Johnston, R. (1989), *Environmental Problems: Nature, Economy and State* (Belhaven: London).

Ledger, D. C. (1981), 'The velocity of the River Tweed and its tributaries', *Freshwater Biology*, 11: 1–10.

Maltby, E. (1986), *Waterlogged Wealth: Why Waste the World's Wet Places?* (Earthscan: London).

Mann, R. H. K. (1988), 'Fish and fisheries of regulated rivers in the UK', *Regulated Rivers: Research and Management*, 2: 411–24.

Mason, B. J. (ed.) (1990), *The surface waters acidification programme* (Cambridge).

Metcalfe, J. L. (1989), 'Biological water quality assessment of running waters based on macroinvertebrate communities: History and present status in Europe', *Environmental Pollution*, 60: 101–39.

Milner, N. (1984), 'Fish', in G. Lewis, and G. Williams (eds.), *Rivers and Wildlife Handbook* (Royal Society for the Protection of Birds/Royal Society for Nature Conservation: Sandy, UK), 51–3.

Moore, P. D., and **Bellamy, D. J.** (1974), *Peatlands* (Elek Science: London).

Moss, B. (1988), *Ecology of Fresh Waters: Man and Medium* (2nd edn.) (Blackwell: Oxford).

National Rivers Authority (1990), *Toxic Blue-Green Algae*, Water Quality Series, 2 (NRA: Peterborough, UK).

Newson, M. D. (1992), 'Conservation management of peatlands and the drainage threat: Hydrology, politics and the ecologist in the UK, in O. M. Bragg, P. D. Hulme, H. A. P. Ingram, and R. A. Robertson (eds), *Peatland Ecosystems and Man: An Impact Assessment* (Department of Biological Sciences, University of Dundee), 94–103.

—— and **Harrison, J. G.** (1978), *Channel Studies in the Plynlimon Experimental Catchments*, Report 47 (Institute of Hydrology: Wallingford, UK).

Parker, R. (1976), *The Common Stream* (Paladin: St. Albans).

Petts, G. E. (1984), *Impounded Rivers: Perspectives for Ecological Management* (John Wiley & Sons: Chichester, UK).

—— (1988), 'Regulated rivers in the United Kingdom', *Regulated Rivers: Research and Management*, 2: 201–20.

Ratcliffe, D. A., and **Oswald, P. H.** (eds.) (1988), *The Flow Country: The Peatlands of Caithness and Sutherland* (Nature Conservancy Council: Peterborough).

Richardson, S. J., and **Smith, J.** (1977), 'Peat wastage in the East Anglian Fens', *Journal of Soil Science*, 28/3: 485–9.

Schumm, S. A. (1977), *The Fluvial System* (John Wiley & Sons: New York).

—— (1985), 'Patterns of alluvial rivers', *Annual Review of Earth and Planetary Sciences*, 13: 5–27.

Smith, I., and **Lyle, A.** (1978), *Distribution of freshwaters in Great Britain* (Institute of Terrestrial Ecology, UK: Edinburgh).

Smith, R. S., and **Chapman, D. J.** (1988), 'The vegetation of upland mires within conifer plantations in Northumberland, Northern England', *Journal of Applied Ecology*, 25: 579–94.

Stevenson, A. (1992), 'The geography of conservation', in M. D. Newson (ed.), *Managing the Human Impact on the Natural Environment* (Belhaven: London), 37–55.

Stevenson, A. C. (and 15 others) (1991), *The Surface Waters Acidification Project Palaeolimnology Programme: Modern Diatom/Lake Water Chemistry Data-set* (Ensis: London).

Strahler, A. (1957), 'Quantitative analysis of watershed geomorphology', *Transactions of the American Geophysical Union*, 38: 913–20.

Ullman, D. (1979), *Hydrobiology* (Wiley Interscience: Chichester, UK).

Vannote, R. L., Minshall, G. W., Cummins, K. W., Sedell, J. R., and Cushing, G. (1980), 'The river continuum concept', *Canadian Journal of Fisheries and Aquatic Sciences*, 37: 130–7.

Walton, I. (1653), *The Compleat Angler or the Contemplative Man's Recreation* (Richard Marriot: London).

Williams, J. G. (1989), 'Snake River spring and summer chinook salmon: Can they be saved?' *Regulated Rivers: Research and Management*, 4: 17–26.

Wright, J. F., Armitage, P. G., and Furse, M. T. (1989), 'Prediction of invertebrate communities using stream measurements', *Regulated Rivers: Research and Management*, 4: 147–55.

Chapter 8

Acid Waters Review Group (1989), *Acidity in UK Fresh Waters* (HMSO: London).

Ashmore, M. R., Bell, J. N. B., and Garretty, C. (1988), *Acid Rain and Britain's Natural Ecosystems* (Imperial College Centre for Environmental Technology: London).

Bergstrom, L., and Jarvis, N. J. (1991), 'Prediction of nitrate leaching losses from arable land under different fertilization intensities using the SOIL-SOILN models', *Soil Use and Management*, 7/2: 79–85.

Brimblecombe, P. (1987), *The Big Smoke: A History of Air Pollution in London since Medieval Times* (Methuen: London).

Coles, T. F., Southey, J. M., Forbes, I., and Clough, T. (1989), 'River wildlife databases and their value for sensitive environmental management', *Regulated Rivers: Research and Management*, 4/2: 179–89.

Conway, G. R., and Pretty, J. N. (1991), *Unwelcome Harvest: Agriculture and Pollution* (Earthscan: London).

Cook, H. F. (1991), 'Nitrate protection zones: Targeting and land use over an aquifer', *Land Use Policy*, 8/1: 16–28.

Croll, B. T., and Hayes, C. R. (1990), 'Nitrate and water supplies in the United Kingdom', *Environmental Pollution*, 50: 163–87.

Degens, E. T., Kempe, S., and Richey, J. E. (eds.) (1991), *Biogeochemistry of major World Rivers* (John Wiley & Sons: Chichester, UK).

Department of the Environment/National Water Council (1980), *General Principles of Sampling and Accuracy of Results* (HMSO: London).

—— (1986), *Digest of Environmental Protection and Water Statistics* (HMSO: London).

—— /Forestry Commission (1991), *Forests and Surface Water Acidification* (DOE/FC: Reading, UK).

Dutcher, J. S., and Boyd, M. R. (1979), *Biochemical Pharmacology*, 28: 3367.

Ellis, B. (1991), 'Urban runoff quality in the UK: Problems, prospects and procedures', *Applied Geography*, 11: 187–200.

Fewtrell, L. (1991), 'Freshwater recreation: A cause for concern?' *Applied Geography*, 11: 215–26.

Fleming, G. (1988), 'Landfill and water: A difficult mix', in *Hydrology of Landfill Sites* (British Hydrological Society/Scottish Hydrological Group: Glasgow).

Foster, S. D. D., Bridge, L. R., Geake, A. K., Lawrence, A. R., and **Parker, J. M.** (1986), *The Groundwater Nitrate Problem*, Hydrogeological Report 86/2 (British Geological Survey: Wallingford, UK).

Jones, F. (1991), 'Public health aspects of the water cycle: A review', *Applied Geography*, 11: 179–86.

Haigh, N. (1989), *EEC Environmental Policy and Britain* (Longman: Harlow, UK).

Hawkins, K. (1984), *Environment and Enforcement: Regulation and the Social Definition of Pollution* (Oxford).

Henderson-Sellers, B., and **Markland, H. R.** (1987), *Decaying Lakes: The Origin and Control of Cultural Eutrophication* (John Wiley & Sons: Chichester, UK).

Holdgate, M. W. (1979), *A Perspective of Environmental Pollution* (Cambridge).

Kirby, C., Newson, M. D., and **Gilman, K.** (1991), *Plynlimon Research: The First Two Decades*, Report 109 (Institute of Hydrology: Wallingford, UK).

Macklin, M. G. (1992), 'Metal pollution of soils and sediments: A geographical perspective', in M. D. Newson (ed.), *Managing the Human Impact on the Natural Environment* (Belhaven: London), 172–95.

Malm, O., Pfeiffer, W. C., Souza, C. M. M., and **Reuther, R.** (1990), 'Mercury pollution due to gold mining in the Madeira River basin, Brazil', *Ambio*, 19/1: 11–15.

Nature Conservancy Council (1984), *Nature Conservation in Great Britain* (NCC: Shrewsbury, UK).

Newson, M. D. (1991*a*), 'Space, time and pollution control: Geographical principles in UK public policy', *Area* 23/1: 5–10.

—— (1991*b*), 'Catchment control and planning: Emerging patterns of definition, policy and legislation in UK water management', *Land Use Policy*, 8/1: 9–15.

—— (1992), 'Land and water convergence, divergence and progress in UK policy', *Land Use Policy*, 9/2: 111–21.

Ormerod, S. J., and **Tyler, S. J.** (1989), 'Long-term change in the suitability of Welsh streams for dippers *Clincus clincus* as a result of acidification and recovery: A modelling study', *Environmental Pollution*, 62: 171–82.

Petts, G. E. (1988), 'Regulated rivers in the United Kingdom', *Regulated Rivers: Research and Management*, 2: 201–20.

Macklin, M. G. (1992), 'Metal pollution of soils and sediments: A geographical perspective', in M. D. Newson (ed.), *Managing the Human Impact on the Natural Environment* (Belhaven: London), 172–95.

Rees, J. (1989), *Water Privatization*, Research Papers, (Department of Geography, London School of Economics: London).

Renberg, I., and **Battarbee, R. W.** (1990), 'The SWAP Palaeolimnology Programme: A synthesis', in B. J. Mason (ed.), *The Surface Waters Acidification Programme* (Cambridge), 281–300.

Richardson, S. J., and **Smith, J.** (1977), 'Peat wastage in the East Anglian Fens', *Journal of Soil Science*, 28/3: 485–9.

Rivett, M. O., Lerner, D. N., and **Lloyd, J. W.** (1990), 'Chlorinated solvents in UK aquifers', *Journal of Institution of Water and Environmental Management*, 4: 242–50.

Royal Commission on Environmental Pollution (1988), 12th Report, *Best Practicable Environmental Option* (HMSO: London).

Royal Society (1983), *The Nitrogen Cycle of the United Kingdom: A Study Group Report* (Royal Society: London).

Skeat, W. O. (1969), *Manual of British Water Engineering Practice*, iii. *Water quality and treatment* (Institution of Water Engineers: Heffer, Cambridge).

Tebbutt, T. H. Y. (1983), *Principles of Water Quality Control* (Pergamon: Oxford).

Timbrell, J. A. (1989), *Introduction to Toxicology* (Taylor and Francis: London).

Tivy, J. (1987), 'Nutrient cycling in agro-ecosystems', *Applied Geography*, 7: 93–113.

Wadsworth, G. A., and **Webber, J.** (1980), *Deposition of Minerals and Elements in Rainfall*, Reference Book 326 (Ministry of Agriculture, Fisheries, and Food, HMSO: London), 47–55.

Waters, D., Jenkins, A., Staples, T., and **Donald, A. P.** (1991), 'The importance of hydrological source areas on terrestrial liming', *Journal of the Institution of Water and Environmental Management*, 5: 336–41.

Chapter 9

Arnell, N. W., Brown, R. P. C., and **Reynard, N. S.** (1990), *Impact of Climatic Variability and Change on River Flow Regimes in the UK*, Institute of Hydrology, Report No. 107, 15 pp. (Wallingford).

Askew, A. J. (1991), 'Climate and water: A call for international action', *Hydrological Sciences Journal*, 36/4: 391–404.

Beven, K. (1985), 'Distributed models', in M. G. Anderson, and T. P. Burt (eds.), *Hydrological Forecasting* (John Wiley & Sons: Chichester, UK), 405–35.

Binder, W. (1979), *Grundzuge der Gewasserpflege* (Landesamt für Wasserwirtschaft: Munich, Germany).

—— and **Grobmaier, W.** (1978), 'Bach- und Flusselaufe-Ihre Gestalt und Pflege', *Garten und Landschaft*, 1: 25–30.

—— **Jurging, P.,** and **Karl, J.** (1983), 'Natural river engineering, characteristics and limitations', *Garten und Landschaft*, 1: 25–30.

Blackie, J. R., and **Eeles, C. W. O.** (1985), 'Lumped catchment models', in M. G. Anderson, and T. P. Burt (eds.), *Hydrological Forecasting* (John Wiley & Sons: Chichester, UK), 311–45.

Broer, G. J. A. A. (1991), *Alarm System for Accidental Pollution of the River Rhine*, Publication 201 (International Association of Hydrological Sciences), 329–36.

Brookes, A. (1984), *Recommendations Bearing on the Sinuosity of Danish Stream Channels: Consequences of Realignment, Spatial Extent of Natural Channels, Processes and Techniques of Natural and Induced Restoration*, Technical Report 1 (Freshwater Laboratory, National Agency of Environmental Protection: Silkeborg, Denmark).

—— (1987), 'River channel adjustments downstream from channelization works in England and Wales', *Earth Surface Processes and Landforms*, 12: 337–51.

—— (1992), 'Recovery and restoration of some engineered British river channels', in P. J. Boon, P. Calow, and G. E. Petts (eds.), *River Conservation and Management*, (John Wiley & Sons: Chichester, UK), 337–52.

Falkenmark, M. (1986), 'Fresh water: Time for a modified approach', *Ambio*, 15/4: 192–200.

George, R. J. (1990), 'Reclaiming sandplain seeps by intercepting perched groundwater with eucalypts', *Land Degradation and Rehabilitation*, 2: 13–25.

Glacken, C. J. (1956), 'Changing ideas of the habitable world', in W. L. Thomas (ed.), *Man's Role in Changing the Face of the Earth* (University of Chicago Press), 70–92.

Goldsmith, E., and **Hildyard, N.** (eds.) (1984), *The Social and Environmental Effects of Large Dams*, i. *Overview*; (1986) ii. *Case Studies* (Wadebridge Ecological Centre: Cornwall, UK).

—— and **Trussell, D.** (1992), *The Social and Environmental Effects of Large Dams*, iii. *A Review of the Literature* (Wadebridge Ecological Centre: Cornwall, UK).

Gore, J. A. (1985), *The Restoration of Rivers and Streams: Theories and Experience* (Ann Arbor Science: Stoneham, Mich.).

Keller, E. A. (1975), 'Channelization: A search for a better way', *Geology*, 3: 246–8.

—— and **Hoffman, E. K.** (1976), 'Sensible alternatives to stream channelization', *Public Works*, 70–2.

Kirkby, M. J., Naden, P. S., Burt, T. P., and **Butcher, D. P.** (1987; 2nd edn., 1993), *Computer Simulation in Physical Geography* (John Wiley & Sons: Chichester, UK).

Lelek, A., and **Kohler, C.** (1990), 'Restoration of fish communities of the Rhine river two years after a heavy pollution wave', *Regulated Rivers, Research and Management*, 5: 57–66.

McConnell, C. A., Parsons, D. R., Montgomery, G. L., and **Gainer, W. L.,** (1980), 'Stream renovation alternatives: The Wolf River story', *Journal of Soil and Water Conservation*, 35: 17–20.

Marchand, M., and **Toornstra, F. H.** (1986), *Ecological Guidelines for River Basin Development* (Centrum voor Milienkunde, Dept. 28, Rijksuniversiteit: Leiden).

Newson, M. D. (1985), 'Twenty years of catchment process studies: What have we taught the civil engineer?', in T. H. Y. Tebbutt (ed.), *Advances in Water Engineering* (Elsevier: London), 39–46.

Nunally, N. R. (1978), 'Stream renovation: an alternative to channelization', *Environmental Management*, 2: 403–11.

Parrish, J. D., Maclolek, J. A., Tunboll, A. S., Hathaway, C. B., and **Norton, S. E.** (1978), *Stream Channel Modification in Hawaii*, Summary Report FWS/OBS/78/19 (Fish and Wildlife Service, US Department of the Interior: Washington, DC).

Petersen, R. C., Petersen, L. B-M., and **Lacoursiere, J.** (1992), 'A building-block model for stream restoration', in P. J. Boon, P. Calow, and G. E. Petts (eds.), *River Conservation and Management* (John Wiley & Sons: Chichester, UK), 293–309.

Prowse, C. W., Yazdani, S., and **Larson, T.** (1986), RUNMOD (software) (John Wiley & Sons: Chichester, UK).

Schofield, N. J. (1990), 'Determining reforestation area and distribution for salinity control', *Hydrological Sciences Journal*, 35/1: 1–19.

Shiklamanov, I. A. (1989), 'Climate and water resources', *Hydrological Sciences Journal*, 34/5: 495–529.

Troendle, C. A. (1985), 'Variable source area models', in M. G. Anderson, and T. P. Burt (eds.), *Hydrological Forecasting* (John Wiley & Sons: Chichester, UK), 347–403.

Index

abandoned river channel 122
acid rain 179
acidification 125, 155, 166
  and land use 179
acrotelm 148
air pollution 179
albedo 31
algal blooms 125, 176, 179, 181
alluvial channel 55
Alzheimer's disease 170
anisotropic 45
annual exceedance series 71
annual hydrograph 38
annual series 70
aquiclude 29
aquifer 29–30
  artificial recharge 43
  confined 42
  management 43
  pollution 44
  unconfined 29, 42
Aral Sea 112
artesian pressure 42
Aswan Dam 110
auger hole method 46–7

background quality 165
bankfull discharge 55
baseflow 38, 42
  generation 59
  index 42, 85
bedding planes 42
Best Practicable Means (BPM) 168
Biochemical Oxygen Demand (BOD) 172
bioconcentration 139
biological indicator 135
biological survey 124, 140
biotic assessment 128–9
borehole 30
boundary layer 18
Brahmaputra flood 89

calibration curve 24
canopy:
  precipitation measurements 18–20
  processes in crops 103
catastrophism 64
catchment 10
  calibration 11
  distributed 38, 186
  lumped 38, 56–7, 186
  research 13
  ungauged 75
catotelm 148
channel 55–6
  classification 54, 124–5
  cross-section 24
  formation 54–5
channelization 113, 141, 144, 191

chemical dilution 24
climatic change 9
  in Britain 197
  effects 94
  in semi-arid areas 197–8
climatic patterns 6
compensation water 83
confidence limits 185
conservation 190, 195
  status 144
  value 126
constant of channel maintenance 53
contamination 156
convection 15
cost–benefit analysis 77, 192
cost–benefit ratio 74
Coto de Doñana 154
critical loads 179
current meter 24
cylinder infiltrometer 46

dam 82
  failure 76, 89
  impacts 113–14, 183
  proofing 76
Darcy's Law 40–1
data-quality control 34
Decision Support Systems (DSS) 192
desertification 81, 94
design lifetime 73–4
diatom records 179
diplotelmic (peat) 148
dipwell 29
drainage:
  arterial 113
  density 53, 124
  impacts 102
  mole 49
  soil 48
  urban 52, 106
drought 82
  causes 81
  episodic 79
  as a hazard 65
  impacts 82
  prediction 84–5, 198
  protection 83
dry deposition 155
dynamic contributing area (DCA) 59
dynamic equilibrium 66

Earth Summit 193
EC policy 168–9
ecology principles 119
ecosystem 3
  freshwater 92, 120
  threats to 110
ecotones 120
ecotoxicology 139

electro-fishing 136
emission standards 169
energy balance 31
environmental assessment 77, 94
environmental hydrology 38, 92
Environmental Impact Assessment (EIA) 192
    of channelization 141
    of dams 114
environmental indicators 138
environmental management 57, 192
    global 184
    sustainable 192
environmental science 3
environmental stress 139
    of water transfer 141
epidemiology 161
epilimnion 130
euphotic zone 129–30
eutrophic waters 146, 182
eutrophication 182
    cultural 178
    effects 178
    problems 179
evaporation 30–1
    actual 32–3
    global rates 9
    measurements 36
    pans 31
    potential 32–3
    rates 11
    of snow 17
evaporimeter 32
evapotranspiration 31

fens 126, 146
field capacity 45
fisheries 135
fissure 42
fissure flow 39
flood 56
    control 78
    data series 71
    duration 72
    as a hazard 65
    hydrograph 12, 26, 67–70
    intensifying conditions 68
    mean annual 72
    measurement 68, 87–89
    peak 69
    prediction 72
    protection 71, 73–4, 76
    warning 77
flood season 80
    climate classification 85
Flood Studies Report 76
flood warning-system 15
flow-gauging 13, 36
flume 24–5
forecast 185
forest-edge effect 149
forest management 97
    runoff 61
    spatial scale 97
    tropical 62–3
forest plantation 94, 101
    cropping cycle 95
freshwater classification 127
freshwater resources 6

frictional resistance 23
    surface flow 52
frontal activity 15

Gabëikovo–Nagymaros scheme 115
gauging station 26
General Circulation Models (GCMs) 197
General Systems Theory 3
Generalized Extreme Value distribution see
    Gumbel distribution
geographical information systems (GIS) 110
global atmospheric system 8
Global Energy and Water Cycle Experiment 9
Global Environment Water Cycle Experiment
    (GEWEX) 196
global water cycle 3
    human impacts 9
    human modification 4
    water balance 6, 8
greenhouse effect 9, 66
ground surface detention 50
groundwater 12, 38
    catchment 30
    and drought 80
    exploitation 43, 80
    monitoring 13
    mound 148
    pollution 39, 44, 175
    recharge 42, 44
    storage 29
Gumbel distribution 72

habitat 27, 119
    classification 128–9, 133
    lentic 132
    lotic 132
hazard 64–5
Holmes's classification of channel vegetation 144
Hortonian overland flow 51–2
humification 148
hydraulic civilization 3, 157
hydraulic conductivity 40
    of peat 47
hydraulic geometry 55
hydroclimatology 14
    of floods 73, 87
hydrograph 74–6
hydrological cycle 3, 110
    forest impact on 94–5
hydrological index 11
hydrological measurements 10
    problems of 36–7
    scale of 11
hydrological models 75
hydrometric measurements 112
hydrometric networks 13, 34, 151–2, 195
hydropolitics 193
hydrosere 130, 144
hypolimnion 130
hyporheic flow 21, 57
hyporheic zone 42

impoundments 113–14
infiltration rate 45
    measurement 46
    in urban areas 53, 108
instantaneous flood peaks 69
Instream Flow Incremental Methodology 133

integrated pollution control 170–1
interactive system 184
interception 20, 97–8
   of intermediate crops 149
   loss 101
   of tropical forests 98–9
International Hydrological Decade 6, 13
ionic balance 130
irrigated agriculture 89
irrigation 49
   canal 84
   design 84–105
   spray 103
   sprinkler 83
   system 9
   trickle 83
   water demand 111
   water losses 50, 195
   water quality 83
   and wetlands 50, 154
isotropic medium 45

Khartoum flood-wave 87
kick sampling 135
Köppen climate classification 36

lake:
   acidified 179
   pollution 164
   stratification 129
   thermal stratification 130
   water quality 130
laminar flow 52
land management 90, 176
land use 90, 176
   effects of 109–10
landfill 175
landscape equilibrium 122
leachate 175
low flow 82, 84–5
lysimeter 33
   forest 95

macrophytes 133–4
macropores 45, 47
MAGIC 189
Manning equation 21, 51, 68
matrix potential 45
Maximum Allowable Concentrations (MACs) 169
mean annual flood 72
Mercury 182–3
mesotrophic 146
Mikri Prespa wetland 154
mires 153
   raised 153
models:
   deterministic 185
   distributed 186, 188
   global atmospheric 8
   heuristic value 189
   hydrological 185
   lumped 186, 189
   parameters 186
modulation 120
monitoring 165, 193
   retrospective 165
Murphy's Law 42

National Rivers Authority 120
nature conservation 119
neutron probe 29
Nile 110, 193
nitrate pollution 176–8
Nitrate Sensitive Areas (NSAs) 178, 182
nival floods 87

oligotrophic 146, 182
orographic enhancement 15
out-of-bank flow 21
overland flow 51

Partial Contributing Area (PCA) 57–8
Partial Duration Series 70
peak-discharge rate 76
peak-flood flow 21
   measurement 68–9
peat extraction 151, 154
   formation 148
Penman 32
percolation 39
permeability 39, 42
   soil 45
phreatophyte vegetation 100
phytoplankton 133
Piché evaporimeter 32
piezometer 47
plough layer 102
policy development 192
pollution 156
   developments 161
   diffuse source 162, 164–5, 176, 179
   impacts 5
   index 134
   monitoring 165
   point source 140, 162, 164, 172, 176
   risk 82
   space/time framework 155
   surveying 166
pollution control 119, 166–7
   in the UK 167–8
   in the EC 168
porosity 39, 42
   soil 45
potentiometric surface 42
Precautionary Principle 168, 185
precipitation 13–14
   measurement 11, 36
prediction 101
   deterministic 185
   statistical 185
pressure transducer 26
probability 73
Probable Maximum Flood 76
purification, natural 157, 160
   of industrial waste 173
   of water 170

quickflow 58–9
   and snowmelt 61

rainfall distribution 10
   variability 36, 85
rainfall simulator 46
raingauge 11
   ground-level 15
   network 12, 15

plastic sheet 20
  recording 15
  standard 15
  storage 14
  tipping-bucket 15
rating curve *see* calibration curve
rational formula 107
recycling processes 157
regime equation 55
regulatory action 165
remote sensing 36, 183, 185, 198
reservoir regulation 83, 85
residence time 5
return surface flow 52
Rhine Action Programme 196
riparian management 144
riparian rights 83, 90, 161
riparian zone 57
risk 73
river basin 11, 189
  conservation 190
  management 122, 189
  preservation 190
river continuum 128
river corridor 57, 129
River Invertebrate Prediction and Classification
    Scheme (RIVPACS) 135
river regulation 125
  impacts 114–15, 140–1
river training 141
root constant 33
RUNMOD 186
runoff:
  artificial 91–2
  components 38
  distribution 10
  domains 38
  impacts 102
  processes 12
  season 26
  urban 105, 108
  variation 60–3
runoff coefficient 11, 27
  roof 109

saline intrusion 43
sanitary age 160
seasonal runoff period 12
sediment transport 55
sensor and recording systems 34
sewage disposal 157
sewage treatment 170, 172–3
  works 156, 160, 172
sewerage 105, 156
sheet flow 52
sludge 171
slurry lagoons 173
snow measurement 17
snow-pillow technique 17
snow-water equivalent 17, 60
soil:
  horizons 7
  pipes 59
  salinized 83
  saturation 29
soil-moisture 44–5
  accounting 49
  content 28

deficit 84
  measurement 27, 29
  sub-system 12
  and vegetation patterns 48
specific yield 39
spillway 76
statistical distribution 71
stochastic modelling 185
storage 3, 5, 12, 27
storm drainage 105
strand line 68
stream networks 53
stream-ordering 124
stream restoration 195
streamflow 11
  measurement 20–1
sulphur dioxide 179
surface-flow generation model 51
surface roughness 102
surface-water acidification 155
sustainable development 193
systematic approach 3

tapping pipe 25
tensiometer 45
Theissen polygon 15
thermocline 129
threshold 66
  behaviour 93
  point 139
throughfall 18, 20
toxicology 161
transpiration 97
trash marks *see* strand line
trophic status 128

unit hydrograph 75
urban channel 109
  drainage design 106
urban hydrology 52, 107
  scale 107
urban water supply 4, 105, 160

vapour pressure deficit 31
vegetation assessment 135
velocity network 55
velocity profile 22–3
volumetric gauging 23
von Post scale 148

Wadi Dhamad 36
wadis 62
waste:
  agricultural 173
  industrial 173
water abstraction 110–11
water balance:
  catchment 12–13
  components 11
  global 10, 36
  in humid temperate zone 36
  regional 3
water pollution 90
water quality:
  classification 125
  decline 43
  studies 18
water requirements 9
water resources 90

distribution 9
exploitation 122
water scarcity 81
water-table 30
  fluctuations 43
  in unconfined aquifers 42
  in wetlands 144
water transfer 141
water use:
  biological 3
  crop 105
  forest 98
  human 4, 11
watershed 10
weir 124

wetlands 78, 124
  classification 146
  damage to 126
  drainage 149–51
  extent 144
  management 153
  soils 148
  tropical 154
  value of 153
wilting point 45
World Climate Programme 196
World Climatic Research Programme 9

zonation 128
zone model 128